AF553429

MARKETING PRACTICES OF DAIRY UNITS

MARKETING PRACTICES OF DAIRY UNITS

By

Dr. Bhooma Parameswara Reddy

Department of Commerce
Sri Venkateswara University
Tirupati
Andhra Pradesh
(India)

DISCOVERY PUBLISHING HOUSE PVT. LTD.
NEW DELHI-110 002

Published by:
Tilak Wasan

DISCOVERY PUBLISHING HOUSE PVT. LTD.
4831/24, Ansari Road, Prahlad Street
Darya Ganj, New Delhi-110002 (India)
Phone: +91-11-23279245, 43764432
Fax: +91-11-23253475
E-mail: parul.wasan@gmail.com
discoverypublishinghouse@gmail.com
info@discoverypublishinggroup.com
web: www.discoverypublishinggroup.com

First Edition: 2011
ISBN: 978-81-8356-797-8

Marketing Practices of Dairy Units

Printed at:
Shree Balaji Art Press
Delhi

Foreword

I am indeed delighted to know that Dr. B. Parameswara Reddy has written book on the topic of *Marketing Practices of Dairy Units.* The author has presented the work systematically with incisive analysis. It is true that dairy industry, of late, has been developing rapidly providing employment to several unemployed in the rural India. The topic is very relevant to the present-day situation.

The work has been presented in nine chapters. Dr. Reddy has shown a sense of commitment in each chapter and contents of each chapter justify the title of the chapter. He presented lucidly the state of dairy industry in the country, in A.P., and in Kurnool district. Each chapter is well supported by both qualitative and quantitative data collected from different sources. He has analysed the marketing practices adopted in the sample dairies in the study area. The analysis has been well supported by tabular form of presentation followed by critical interpretation. Dr. Reddy critically evaluated the marketing strategies of sample units with a focus on salient factors like product line packaging and labelling and branding of milk products, product line strategies, sales trends and staple products, product distribution etc. Each variable has been critically analysed for sample organizations. Further, he has also analysed the pricing policies and promotion strategies in the sample organizations at length. In a nutshell, he has made sincere attempt in testing the marketing practices adopted in the sample units and analysed the relative strengths and weaknesses of the same in the respective organizations. In line with the analysis and interpretation, he has drawn logical inferences and offered some useful suggestions which will be a concrete feed back to the organizations and for framing effective marketing strategies. This study will be definitely useful for not only to the dairy industries but also to dairy economics, commerce and trade disciplines.

Prof. J.V. Prabhakara Rao,
Vice-Chancellor,
Rayalaseema University,
Kurnool-518002,
Andhra Pradesh

Preface

Marketing is one of the most important intervening variables, if not causal one for organizational effectiveness, it is because all outputs in the form of products have to ultimately reach from producers to consumers. As such importance of marketing for any organization cannot be ignored. Now-a-days marketing is playing a predominant role in the process of industrialization which in turn emerged as a concept for the economic development of a country.

In ancient and middle ages the place of cattle in the economic and social life of people gained importance. By the time Britishers came to advent India; the little republics (villages) were flooded with milch animals, occupying a unique position among the world nations. India has emerged as the largest world producer of milk with an annual production of 100 million tonnes in 2007, surpassing the production of 88 million tonnes by U.S.A. in the same year. The milk production in the country quadrupled from 23 million tonnes in 1973 to 100 million tonnes in 2007, with the remarkable annual growth rate of 4.5 per cent as against the world's average of about one per cent. The per capita availability of milk was 245 grams per day in 2007, which was higher than the requirement of 220 grams per day as recommended by Indian Council of Market and Research (ICMR) 2007. Consequent to the New Economic Policy, 1991 and the recent amendments to the Milk and Milk Products Order (MMPO)1992. India offers a level playing field to Indian and foreign investors alike to invest in dairying both with a view to serve domestic and export markets. India received further boost with the signing of Uruguay Round

Agreement on Agriculture (URAOA) in 1994, eventually culminating in the establishment of the World Trade Organization (WTO) in 1995. The dairy industry was delicenced in 1991 and the private sector, including Multi National Companies (MNCs), were facilitated to set up milk processing and product manufacturing plants.

The organized dairying in India was began at the end of 19th century when military dairy farms and creameries were started to meet the demands of the armed forces and their families. The first dairy co-operative society was established at Allahabad in 1913. After this the Calcutta milk supply societies union was established in 1919. The first major landmark in the development of co-operative dairying in India was the establishment of the Kaira District Co-operative Milk Producers Union Ltd. (KDCMPUL). Before the establishment of Amul, the milk marketing system in the district was controlled by contractors and middle men who exploited the milk producers in all possible ways, thereby earning huge profits. As a result there was growing discontent among the milk producers. On the advice of Sri Sardar Vallabhai Patel, and under the able leadership of Sri Thribhuvandas Patel, it was decided to organize primary dairy cooperative societies which led to the emergence of Amul on 14th December, 1946.

The dairy co-operatives in Kaira district under the able guidance of Sri Thribhuvandas Patel and Dr. Varghese Kurien, its chairman and general manager respectively, followed an integrated approach to dairy development linking all the major elements of dairying viz., procurement, production, processing and marketing and achieved remarkable progress. Today Amul is the largest dairy plant in the country handling on an average about 8.5 lakh liters of milk per day collected from 3.65 lakh milk producers from over 880 villages of the district. The ultimate result is that Anand or Amul became a model world-wide for a co-operative. Amul also started a value loaded milk products like butter, cheese etc. The spectacular growth of dairy sector in the country was mainly due to the implementation of 'Operation Flood' programme. The three phases of Operation Flood, OF-I (1970-79), OF-II (1979-85), OF-III (1986-94) were for the largest externally aided dairy development projects in the world. National Dairy Development Board (NDDB) established in 1965 by Shri Lal Bhahadur Sastri, Prime Minister of India, was entrusted with the responsibility of implementation of the Operation Flood programme. Its aim has been to replicate Anand pattern dairy cooperatives throughout the country. Specific importance is also given for the improvement of the germ plasm of the cattle, better animal health care and nutritious feed and fodder. These measures resemble pre-harvest packages in agriculture, market support, storage facilities, establishment of distribution network and marketing policies and strategies are very vital for the development of dairy industry on sound lines.

The largest institutional arrangements such as collection, handling, transportation, processing, packaging and retailing the share of the organized sector in milk marketing is 40 per cent. However, the cooperative arrangements for liquid milk marketing have helped in commercializing the Indian dairy sector. The corporate sector has been increasing its share in marketing in the area of high value milk products. Because, marketing is an integrated net work comprising all these agents, the above mentioned design is formulated. Especially in the area of marketing, the functions like procurement, storage, transportation and packaging are comprehensively studied for an analytical and critical appraisal of the marketing practices in the select dairy units.

The book is divided into nine chapters. First chapter is devoted to introduction. In this chapter significance of dairy farming, origin, development, global dairy scenario, cooperative dairying, operation floods and fodder development are dealt with in Part-A. Part-B is devoted for the dairy development in Andhra Pradesh. The second chapter consists of Part A and Part B. Part-A deals with the review of literature, objectives and research methodology of the study is covered in Part B. The third chapter gives exhaustively information about Vijaya and Nandi dairies profile in Kurnool and Rayalaseema districts. Milk procurement, pricing, policies, sales and channels of distributions are discussed in detail here. In the fourth chapter I have tried to analyse various issues relating to procurement of milk in terms of quantity and spectral variations focused on selected dairies. Product related marketing practices of Vijaya and Nandi dairies are highlighted in chapter five. The sixth chapter dealt with pricing objectives, demand estimation methods, pricing methods and price discounts and allowances relating to the sample dairies. Promotion practices in sample dairy units are presented in chapter seven. Advertisement programmes and budgets, sales and distribution practices of dairies, channels of distribution, push-pull strategies, retail outlets and dealers and direct sales are studied here in the penultimate chapter i.e. 8th chapter. Summary of findings, conclusions, suggestions and area's of further research are dealt with the chapter nine.

Dr. Bhooma Parameswara Reddy

Acknowledgements

I wish on record, my deep sense of gratitude to *Prof. J.V. Prabhakara Rao,* Vice-Chancellor, Rayalaseema University, Kurnool, Andhra Pradesh, for his constructive criticism and expert advice.

I wish to express my sincere thanks to *Prof. B. Bhagavan Reddy,* Head of the Department of Commerce, Sri Venkateswara University, Tirupati, for their valuable suggestions.

I am deeply indebted for ever to *Prof. S. V. Subba Reddy,* Associate Professor, DDE, Tirupati for his guidance, encouragement and help. As a guide and supervisor he not only gave me an opportunity to undertake this study, but also inspired me throughout the study.

I am very much indebted to, *Prof. M. Munirami Reddy, Prof. C. Sivarami Reddy, Prof. P. Mohan Reddy, Prof. B. Ramachandra Reddy,* and other faculty members for their assiduous assistance and cooperation.

I owe my thanks to Sri *P. Srinivasa Babu,* Marketing Executive, Vijaya Dairy and Sri *A Prasada Reddy,* Marketing Manager, Nandi Dairy, Nandyal, Kurnool district for their cooperation and provide useful information for this research work.

I am highly grateful to my brothers *B. Rameswara Reddy* and *B. Jagadheswara Reddy* and other of my family members for their affectionate blessings to complete of my research work.

I would like to thanks to *Shri Tilak Wasan and Shri Parul Wasan,* Discovery Publishing House Pvt. Ltd., New Delhi, for published of my book.

Dr. Bhooma Parameswara Reddy

Contents

Foreword

Preface

Acknowledgements

Abbreviations

1. **Introduction : An Overview of Dairy Development in India and Andhra Pradesh 1**

Introduction—Part A : Dairy Development in India—Significance of Dairy Farming in India—Growth of Crossbreed Cows vis-à-vis Indigenous Cows—Inter-State Variations in Milk Production and Yield—Global Diary Scenario—Steady Growth of Dairy Sector—Co-operative Dairying—Origin—Development—Anand Pattern—The Replication of Anand Pattern—Operation Flood Programmes—Operation Flood-I (1970-81)—Operation Flood-II (1981-85)—Operation Flood-III (1985-2002)—Present Status of Operation Flood Programme—Dairy Development during Plan Periods—Fodder Development—Dairy Development during the First Two Decades of Planning 1951-71—Technology Mission on Dairy Development—Key Village Scheme—Intensive Cattle Development Project—Establishment of Cattle Colonies and Milk Schemes—Dairying in Post-Reform Period—Introduction of Milk and Milk Products Order—Uruguay Round Agreement on Agriculture—Perspective 2010—Strengthening Co-operative Business—Production Enhancement—Quality Assurance Programmes—Information and

Development Research—The Challenges Ahead in Dairy Sector in the Country—Responses to Meet the Challenges—Enhancement of Productivity through Breed Improvement—Milk Consumption Pattern in India vis-à-vis other Countries—Prospects of Dairy Industry in India—Part-B : Dairy Development in Andhra Pradesh—Significance of Dairying in Andhra Pradesh—A.P. State Domestic Product and Its Sectoral Shares—Growth and Structure of Dairying A.P.—Formation of the Andhra Pradesh Dairy Development Corporation (APDDC)—Objectives and Role of APDDC—Formation of Dairy Development Co-operative Federation (APDDCF)—Objectives of the Federation—Membership of Federation—Dairy Development during Plan Periods in A.P. —Second Five-year Plan (1956-61)—Third Five-year Plan (1961-66)—Fourth Five-year Plan (1969-74)—Fifth Five-year Plan (1974-78)—Sixth Five-year Plan (1980-85)—Seventh Five-year Plan (1985-90)—Annual Plans (1990-92)—Eighth Five-year Plan (1992-97)—Dairy Development Under OF—Operation Flood-I—Operation Flood-II—Operation Flood III—Development of Dairy Infrastructure in the State—Bovine Population in A.P.—Milk Production and Per Capita Availability of Milk—Procurement and Sale of Milk—Conclusion—References.

2. Review of Literature and Research Design 63

Introduction—Part-A : Review of Literature—Studies on Marketing of Dairy Products—Studies on Economics of Dairying—Studies on Women Participation in Dairying—Studies on Bovine Population and Milk Yields—Studies on the Consumption and Availability of Milk—Studies on Capacity Utilization and Production—Studies on Cost Structure and Returns in Dairying—Part-B : Objectives and Research Methoidology of the

Study—Statement of the Problem—Need for and Significance of the Study—Objectives of the Study—Sample Design —Data Sources—Tools of Analysis—Scope and Limitations—Organisation of the Book—References.

3. Profiles of Sample Dairies 87

Introduction—Geographical Location of Kurnool District—Climate Conditions and Revenue Administration Set-up —Infrastructure for Dairy Development—Part-A : Profile of Vijaya Dairy—Origin of KDMPMACU—Services of KDMPMACU Ltd.—Fertility Campaigns—Vaccination—Provision of De-worming Drugs—Insurance—Feed—Urea Treatment—Others Services—Milk Procurement and Prices are Discussed in this Section—Contents and Shelf Life of Milk and Milk Products—Price Fat and SNF Contents and Shelf Life of Milk Products and Milk by Products of Vijaya Dairy—Sales of Milk, Milk Products and By-products in Vijay Dairy—Part-B : Profile of Nandi Dairy—Beginning of Nandi Group ofgggg Industries—Origin and Growth of the Nandi Dairy—Milk Procurement by Nandi Dairy—Procurement Prices, Processing and Sales in Nandi Dairy—Channels of Distribution of Nandi Dairy—Sales of Milk Products, By-products and Products Prices of Nandi Dairy —Organization Chart of Nandi Dairy—Conclusion.

4. Trends in Milk Procurement in Sample Dairies 116

Introduction—Milk Procurement and Production in Vijaya Dairy—Trends in Milk Procurement and Production in Nandi Dairy—Lean and Peak Seasonal Variations in Milk Procurement in Vijaya Dairy—Lean and Peak Seasonal Fluctuations in Milk Procurement in Nandi Dairy—Capacity Utilization in Milk Chilling Centers of Vijaya and Nandi Dairies—Conclusion.

5. Product Related Marketing Practices 128

Introduction—Product of Sample Dairy Units—Product Hierarchy—Product-line Sales in Sample Dairy—Product Depths in Samples Dairies—Stretching of Packaging and Labelling in Sample Dairies—Ingredient Branding—UP - and Down-Stretching of Product Lines—Trend of Sales of Core and Staple Dairy Products—Distribution of Sales by-Product Lines—Conclusion.

6. Pricing Objectives and Methods of Dairies 155

Introduction—Pricing Objectives—Demands Estimations Techniques—Pricing Methods—Price Discount and Allowances—Conclusion.

7. Promotion Practices in Sample Dairy Units 166

Introduction—Advertising Programmes in Sample Dairy Units—Sales Promotion Programmes—Events/Experiences Programs—Public Relations—Personal Selling—Direct Marketing—Rank Order of Communication Mix—Communication Budgets and Sales—Advertising Budgets and Sales—Sales Promotion—Media-wise Advertising Budgets—Conclusion.

8. Sales and Distribution Practices of Dairies 191

Introduction—Marketing Channel Systems and Strategies—Sales through Dairy Owned Retail Outlets—Distribution through Dealer Networks—Instructional vis-à-vis total Sales —Brand Names—Trend of Sales through Different Channels of Distribution—Conclusion.

9. Summary of Findings, Conclusions and Suggestions 205

Part-A : Findings—Part-B : Conclusions—Part-D: Areas for Further Research.

Bibliography 217

Index 223

Abbreviations

AH	:	Animal Husbandry
AI	:	Animal Insemination
AIRCRC	:	All India Rural Credit Review Committee
AMUL	:	Anand Milk Union Limited
APDDC	:	Andhra Pradesh Dairy Development Corporation
APDDCF	:	Andhra Pradesh Dairy Development Cooperative Federation
AP	:	Andhra Pradesh
AWDCS	:	All Women Dairy Cooperative Societies
BO	:	Butter Oil
DAHD	:	Department of Animal Husbandry and Dairying
DCS	:	Dairy Cooperative Societies
DMS	:	Delhi Milk Scheme
DCMPU	:	District Cooperative Milk Producers Union
DMPU	:	District Milk Producers Union
EEC	:	European Economic Committee
FMD	:	Foot and Mouth Disease
gms	:	Grams
ICMR	:	Indian Council of Market and Research
IDA	:	Indian Dairy Association
IMP	:	Integrated Milk Project
IMSS	:	Integrated Milk Supply Scheme

IITM	:	Indian Immunological Technology Mission
KVS	:	Key Villages Scheme
KCDL	:	Katara Cooperative Dairy Limited
Kg.	:	Kilo Grams
KDMPMACU	:	Kurnool District Milk Producers Mutually Aided Cooperative Union
KDCMPU	:	Kaira District Cooperative Milk Producers Union
KDCMPU	:	Kurnool District Cooperative Milk Producers Union
Lpa	:	Liter Per Annum
llpa	:	Lakh litre Per Annum
MNCs	:	Multi National Companies
MMPO	:	Milk and Milk Products Order
MMSU	:	Madras Milk Supply Union
MCS	:	Milk Cooperative Societies
MPMACU	:	Milk Producers Mutually Aided Cooperative Union
MCAs	:	Milk Collection Agents
NCA	:	National Commission on Agriculture
NCDDC	:	National Cooperative Dairy Development
NMG	:	National Milk Grid
OF	:	Operation Flood
SMMF	:	State Milk Marketing Federation
SPMA	:	Sanitary and Physto Sanitary Measures Agreement
TMDD	:	Technology Mission on Dairy Development
UMB	:	Urea Molasses Block
WFP	:	World Food Programme

CHAPTER 1

INTRODUCTION : AN OVERVIEW OF DAIRY DEVELOPMENT IN INDIA

INTRODUCTION

This chapter designed to present evolution of dairy industry in India and Andhra Pradesh (AP). In Part-A discusses broader dimensions of Indian dairy industry such as its share in GDP, growth pattern of livestock population, India milk production *vis-à-vis* other top dairying countries, Anand Pattern of dairy development, dairy development in I, II and III phases of Operation Flood, plan-wise outlays and actual expenses on dairying, WTO and post-reform measures and their impact on dairying. Part-B focuses on dairy development in Andhra Pradesh at state level. This part deals with milk production and per capita milk availability, objectives and functions of Andhra Pradesh Dairy Development Corporation (APDDC), dairy development during Five-year Plans, progress under Operation Flood (OF) programme, growth of cattle and buffalo population, milk production and procurement and milk sales.

PART A: DAIRY DEVELOPMENT IN INDIA

Significance of Dairy Farming in India

In ancient and middle ages the place of cattle in the economic and social life of people gained importance. By the time the Britishers come to advent India; the little republics (villages) were flooded with milch animals, occupying a unique position among the world nations. India has emerged as the largest world producer of milk with an annual production of 100 million tonnes in 2007, surpassing the production of 88 million tonnes by U.S.A. in that year. The milk production in the country quadrupled from 23 million tonnes in 1973 to 100 million tonnes in 2007, with the remarkable annual growth rate of 4.5 per cent as against the world's average of about one per cent. Consequent to the New Economic Policy, 1991 and the recent amendments to the Milk and Milk Products Order (MMPO) 1992. India offers a level playing field to Indian and foreign investors alike to invest in dairying both with a view to serve domestic and export markets. India got further boost with the signing of Uruguay Round Agreement on Agriculture (URAOA) in 1994, eventually culminating in the establishment of the World Trade Organization (WTO) in 1995. The dairy industry was delicenced in 1991 and the private sector, including Multi National Companies (MNCs), was allowed to set up milk processing and product manufacturing plants. As efficiency is the key factor in privatization policy.

The Indian dairy sector owes its success to millions of small producers who have one or two milch animals yielding 3-4 litres of milk per day. Annual milk yield of dairy animal in India is about one-tenth of that achieved in the U.S.A. and about one-fifth of the yield of a grass-fed New Zealand dairy cow. Dairying has increasingly become a part of the state's anti-poverty programme. Organizations like the Small Farmers Development Agency (SFDA) and the Integrated Rural Development Programme (IRDP) give priority to dairy development projects as an instrument for uplifting the economic conditions of the weaker sections of the rural population.

Dairying in India is considered as a sub-system of the farming system, for the milch animals are generally fed with crop residues, agricultural wastes, compound cattle feed and oilseed cakes. The cost of milk production in India is one among the lowest in the world. Dairying in India, through the small herd dairy systems with feeding practices that do not place pressure on land, has significant competitive advantages. The low capital investment and steady returns make dairying a covetous activity among the marginal and small farmers and even the landless who depend for fodder on common grazing and forest lands. India has two per cent of the geographical area of the world and supports about 18 per cent of the world's cattle population, but it contributing only around 14 per cent of the world's milk output.

Dairy farming contributes to prosperity of rural folk in more than one way. The Royal Commission on Agriculture (RCA) observed that the cow and the working bullock carry on their pertinent back the whole structure of the Indian agriculture.

The growth pattern of the livestock production *vis-à-vis* GDP during 1951 and 2003 is furnished in Table: 1.1. It can be observed that the livestock production was Rs.139 billion in the initial year 1986 which increased to Rs.1209 billion in the terminal year 2003, recording 77 per cent increased over 17-year period. Livestock production as percentage GDP has been fluctuations between the lowest of 5.37 per cent in 2003 and the highest of 6.85 per cent in 1993.

Growth of Crossbreed Cows *vis-à-vis* Indigenous Cows

Successive rounds of livestock census have clearly established the speed with which crossbreeding has spread in different parts of the country since its beginning in 1960s. Table 1.2 portrays growth pattern of livestock population in India in the reference period, 1951 to 2003. Adult female cattle percentage of cattle was highest 35.02 per cent in 1951 and lowest at 29.40 per cent in 1961. Adult female buffaloes percentage of buffaloes was highest at 52.05 per cent in 2003 and lowest at 47.46 per cent in 1961 and 1966. Total bovines as percentage of total livestock was highest at 67.86 in 1951 and lowest at 58.37 per cent in 2003. From the preceding analysis it can be concluded that the proportion adult female buffaloes in total buffaloes population has been higher has

Table 1.1. Share of Livestock Sector in GDP

(Rs. in billions)

Year	GDP	Value of Livestock Products	
		Livestock Production	Col: 3 %age of Col: 2
1986	2,338	139	5.95
1987	2,600	156	6.00
1988	2,949	183	6.21
1989	3,527	217	6.15
1990	4,087	275	6.73
1991	4,778	308	6.45
1992	5,528	375	6.78
1993	6,307	432	6.85
1994	7,813	507	6.49
1995	9,170	577	6.29
1996	10,733	650	6.06
1997	12,435	747	6.01
1998	13,901	819	5.89
1999	15,981	911	5.70
2000	17,618	992	5.63
2001	19,030	1,093	5.74
2002	20,910	1,187	5.68
2003	22,495	1,209	5.37

Table 1.2. Growth Pattern of Livestock Population in India: From 1951 to 2003

(Rs. in millions)

Year	Cattle	Adult female cattle	Buffaloes	Adult female buffaloes	Total bovines (2+4)	Total livestock
1951	155.3	54.4 (–35.02)	43.4	21.00 (48.38)	198.70 (67.86)	292.8
1956	158.7	47.30 (29.80)	44.9	21.70 (48.33)	203.60 (66.41)	306.6
1961	175.6	51.00 (29.04)	51.2	24.30 (47.46)	226.80 (67.62)	335.4
1966	176.2	51.80 (29.40)	53	25.40 (47.92)	229.20 (66.61)	344.1
1972	178.3	53.40 (29.95)	57.4	28.60 (49.83)	235.70 (66.69)	353.4
1977	180	54.60 (30.33)	62	31.30 (50.48)	242.00 (65.58)	369
1982	192.5	59.21 (30.76)	69.78	32.50 (46.57)	262.40 (62.51)	419.6
1987	199.7	62.07 (31.08)	75.97	39.13 (51.51)	257.80 (61.91)	445.3
1992	204.6	64.36 (31.46)	84.21	43.81 (52.02)	289.00 (61.33)	470.9
1997	198.88	64.42 (32.39)	89.92	46.77 (52.01)	288.80 (59.50)	485.38
2003	185.18	64.51 (34.84)	97.92	50.97 (52.05)	283.10 (58.37)	484.98

Notes: 1. Figures in parentheses in col.3 is adult female cattle expressed as Percentages of cattle in col. 2
2. Figures in parentheses in col. 5 are adult female buffaloes expressed as percentage of buffaloes in col.4
3. Figures in parentheses in col.6 are total bovines expressed as percentage of total livestock in col.7

Source: GOI, Basic Animal Husbandry Statistics 2004, Department of Animal Husbandry and Dairying, Ministry of Agriculture.

compared to the proportion of adult female cattle to total cattle population. Further proportion bovine population in total livestock has shown consistent down trend

Table 1.3 presents annual average growth rates of cattle, buffalo, bovine and livestock population for five year periods stating from 1951-56 and ending with 1997-2003. It can be observed that adult female buffaloes have witnessed then female cows. Bovine population has grown at lower rates than livestock population.

Table 1.3. Annual Growth Rates of Livestock Population: From 1951-56 to 1997-2003

(percentage)

Five-year periods	Cattle	Adult female cattle	Buffaloes	Adult female buffaloes	Total bovines	Total livestock
1951-56	0.43	-2.76	0.68	0.66	0.49	0.93
1956-61	2.04	1.52	2.66	2.29	2.18	1.81
1961-66	0.07	0.31	0.69	0.89	0.21	0.51
1966-72	0.24	0.61	1.61	2.4	0.56	0.53
1972-77	0.19	0.45	1.55	1.82	53	0.87
1977-82	1.35	1.63	2.39	0.76	1.63	2.6
1982-87	0.74	0.95	1.71	3.78	1.01	1.2
1987-92	0.48	0.73	2.08	2.28	94	1.12
1992-97	-0.56	0.02	1.32	1.32	-0.01	0.61
1997-03	-1.18	0.02	1.43	1.44	-0.33	-0.02

Source: GOI, Basic Animal Husbandry Statistics 2004,Department of Animal Husbandry and Dairying, Ministry of Agriculture.

Inter-State Variations in Milk Production and Yield

As in milk production and availability, there are wide interstate variations in milk yields. The average productivity of local cows is highest in Haryana (4.11 kg/day) followed by Punjab (2.88 kg/day) and Gujarat (2.84 kg/day). For crossbreed cows, it is highest in Punjab (8.36 kg/day) followed by Gujarat (7.96 kg/day) and West Bengal (7.82 kg/day). The average productivity of buffaloes is highest in West Bengal (6.26 kg/day) followed by Haryana (5.64 kg/day) and Punjab (5.62 kg/day).

Global Diary Scenario

In the world about 2450 million peoples are involved in agriculture, out of which probably two-thirds or even three-fourths are completely or partially dependent on livestock farming. Until recently many countries have considered milk too bulky and perishable to make long-distance trade feasible. Therefore, they developed capabilities satisfying domestic liquid milk requirements through domestic dairy industries or depended on milk product imports, or a combination of both. For these very reasons, most dairying nations have a complex mechanism to regulate their dairy industries through interventions, financial supports and physical controls.

Cooperatives dominate dairy industry. In the United Kingdom, all the milk produced by farmers is procured by cooperatives. There are no private sector dairy plants in New Zealand. A total of 90 per cent of the dairies in former West Germany are cooperative, and in Denmark, Netherlands and Sweden the entire dairy industry is organized on cooperative lines. In the U.S.A., 70 per cent of the dairy industry is in cooperative sector. Dairy programmes are subject to significant government participation and regulation than most other domestic agricultural industries in the U.S.A.. There are several laws to encourage dairy cooperatives and protect the interests of the farmers. Table 1.4 presents milk production in India has compared to other top nine milk producing countries of the world for 7-year period, 1998 to 2004. It is heartening to note that India's percentage share in total world milk production has kept upward trend, competeting with U.S.A. for top position. The combined share of top ten countries in milk production covers around 70 per cent throughout the reference period. In the year 1998

Table 1.4. Milk Production in India *vis-à-vis* other Countries: From 1998 to 2004

(In '000 tonnes)

Sl. No.	Country	Years						
		1998	1999	2000	2001	2002	2003	2004
1	India	71300 (16.35)	71900 (16.47)	76490 (17.12)	79490 (17.46)	81855 (–17.72)	84380 (–17.96)	88000 (–18.54)
2	U.S.A.	71377 (16.37)	73807 (16.91)	76004 (17.02)	75608 (–16.6)	77139 (–16.69)	77252 (–16.44)	77220 (–16.27)
3	Russia	33197 (7.63)	33140 (7.59)	31938 (7.15)	33000 (–7.25)	33467 (–7.24)	33300 (–7.09)	33000 (–6.95)
4	Germany	28378 (6.51)	24334 (5.57)	28331 (-6.34)	28191 (–6.19)	27874 (–6.03)	28533 (–6.07)	28100 (–5.92)
5	France	24763 (5.70)	21461 (4.91)	24975 (5.59)	24909 (–5.47)	25254 (–5.46)	24590 (–5.23)	24200 (–5.1)
6	Brazil	21630 (4.97)	12700 (4.97)	22134 (4.95)	22580 (–4.96)	22635 (–4.9)	23000 (–4.9)	23500 (–4.95)
7	China	6621 (1.52)	8460 (1.93)	8420 (1.88)	10255 (–2.25)	12998 (–2.81)	17463 (–3.71)	19000 (–4)
8	United Kingdom	14632 (3.35)	15015 (3.44)	14489 (3.24)	14707 (–3.23)	14869 (–3.21)	15017 (–3.2)	14714 (–3.1)
9	New Zealand	10500 (2.40)	11900 (2.73)	12700 (2.84)	13300 (–2.92)	13900 (–3)	14450 (–3.08)	14500 (–3.06)
10	Ukraine	137838 (3.15)	1172 (2.56)	12658 (2.83)	13444 (–2.95)	14142 (–3.06)	13658 (–2.9)	13200 (–2.78)
11	**Total**	**296136 (67.94)**	**292899 (67.13)**	**308139 (69.01)**	**3188988 (70.06)**	**324133 (70.16)**	**331643 (70.60)**	**335414 (70.67)**
12	Other Countries	139726 (32.05)	143404 (32.86)	138391 (30.99)	136320 (2994)	137795 (29.84)	138081 (29.40)	139200 (29.33)
	Grand Total (11+12)	**4355862 (100.00)**	**436293 (100.00)**	**446530 (100.00)**	**455218 (100.00)**	**461928 (100.00)**	**469724 (100.00)**	**474614 (100.00)**

Note: Figures in parentheses are percentages to respective low totals.

India's milk production was on par with U.S.A., which was ranked first by producing 71.4 mt of milk in that year. India held the second rank in the production of milk to the tune of 71.3 mt in the same year. India overtook the United States in milk production in the year 1999 by producing 74.6 mt of milk, as compared to the 73.84 mt production of milk in U.S.A.. With the volume of milk production of 83 mt in the year 2004, India became number one milk producer in the world.

Steady Growth of Dairy Sector

The green revolution has now reached stagnation. In such a case, there is a need for diversification of crop production system by greater integration of livestock and inland fisheries. Fortunately, various government initiatives through the promotion of dairy cooperative movement under Operation Flood Programme and several other dairy production schemes have resulted in augmenting milk production at an average annual growth of 4.5 per cent. The implementation of Operation Flood (OF) brought a "White Revolution" in India with milk production increasing from 21.2 million tonnes per annum in 1968-69 to 66 million tonnes by 95-96, at the end of project period.

Table 1.5 shows the milk production and per capita milk availability during 1951-2007. It can be observed that there is a steady growth of milk production over the years. Despite inter being the largest milk producer in the world, its per capita milk availability is one of the least in the world. It can be also noted that the per capita availability of milk, which declined during the 1950s and 1970s from 124 grams per day in 1950-51 to 121 grams in 1973-74, increased substantially in 1990s and reached about 235 grams per day in 2002. In the year 2007, the per capita availability of milk was 245gms per day which was higher more against the requirement of 220 grams per day as recommended by Indian Council of Market and Research (ICMR) 2007.

Table: 1.5. Milk Production in India: From 1951 to 2007

Year	Production (in million tonnes)	Yearly %age Change	Per capita availability (Grams /day)
1951	17.0	....	124
1961	20.0	5.26	124
1971	22.0	3.77	112
1981	31.6	8.54	128
1991	53.9	3.33	176
2001	80.6	2.94	220
2002	84.8	5.21	235
2003	86.2	1.65	230
2004	88.1	2.2	231
2005	92.5	4.75	233
2006	97.1	5.00	241
2007	100.0	3.00	245
Avg.	64.48	-	191.58
CGR	13.46	-	7.15

CO-OPERATIVE DAIRYING

An impressive development has taken place as far as dairying as co-operatives are concerned. Strengthening the cooperative business is a thrust area that focuses on expanding and reinforcing the cooperative infrastructure at every level and enhancing market potential through modern dairy plant technology, new product development and innovative marketing. Today, women dairy farmers are encouraged to play a major role. Thus, it has become an important instrument for their empowerment; NDDB is committed to increase women participation by establishing 2,062 women dairy cooperative societies with 90,000 women participants.

Origin

The Katara Co-operative Dairy Limited at Allahabad in Uttar Pradesh is probably the oldest existing dairy organization registered under the Co-operative Societies Act, 1912 into which milk producers were admitted as members. Between 1914 and 1919 seven more societies were formed. In the year 1919, Calcutta Co-operative Society (CCS) was started in 1927. Madras Milk Supply Union came into being with first processing facilities. In 1932, the Lucknow Milk Producers Co-operative Union Limited was established in 1945, Aarey Milk Colony was established by the Bombay government under the Greater Bombay milk scheme. Thereafter, till Independence there was no significant progress in the development of dairy co-operatives.

However, after Independence, the first large scale and systematic break through in dairy co-operatives in India was made in 1948 by the Kaira District Co-operative Milk Producers Union at (KDCMPU) Anand, ultimately, the union came to be known as the Anand Milk Union Limited, abbreviated to "AMUL" which in vernacular means 'highly valuable' or 'beyond all prices'.

Development

The organized dairying in India was started at the end of 19th century when military dairy farms and creameries were started to meet the demands of the armed forces and their families. The first dairy co-operative society was established at Allahabad in 1913. After this the Calcutta Milk Supply Societies Union was established in 1919. Till 1938 there were only 19 unions with 264 primary societies and 11,600 milk producer members. However, dairy co-operatives did not make much headway in the pre-Independence period.

The first major landmark in the development of co-operative dairying in India was the establishment of the Kaira District Co-operative Milk Producers Union Ltd (KDCMPUL). Before the establishment of Amul, the milk marketing system in the district was controlled by contractors and middle men who used to exploit the milk producers in all possible ways, thereby earning huge profits. As a result there was growing discontent among the milk producers. On the advice of Sri Sardar Patel and under the able leadership of Sri Thribhuvandas Patel, it was decided to organize primary dairy cooperative societies which led to the emergence of Amul on 14th December, 1946.

The dairy co-operatives in Kaira district under the able guidance of Shri Thribhuvandas Patel and Dr. Varghese Kurien, its chairman and general manager respectively, followed an integrated approach to dairy development linking all the major elements of dairying viz., production, procurement, processing and marketing and achieved remarkable progress. Today AMUL is the largest dairy plant in the country handling on an average about 8.5 lakh liters of milk per day collected from 3.65 lakh milk producers from over 880 villages of the district. The ultimate result is that Anand or Amul became a model world-wide for a co-operative.

Anand Pattern

The unprecedented success of AMUL stimulated the farmers in other districts of Gujarat to emulate their example. Thus, the integrated approach to co-operative dairy was successfully adopted in Gujarat and later it came to be known as 'Anand-pattern of dairy co-operatives. This Anand model of co-operative structure was built on vertically integrated co-operatives linking rural producers with urban consumers. It is a three-tier structure. The three tiers are:

I. The primary Dairy Cooperative Society at the village level is the first-tier that consists of members who own milch animals within the village jurisdiction and supply milk to the co-operative society on regular basis.

II. District cooperative milk producers' union at the district level is the second-tier that in which all primary societies are members. It is managed by a Board of Directors the majority of who are elected by the presidents of primary diary cooperative societies. The district union is responsible for the procurement, processing and marketing of milk and also to provide technical inputs like first-aid, emergency veterinary services, Artificial Insemination (AI) facilities, fodder seeds, cattle feed and for training the staff of the primary societies.

III. Cooperative federation at the State level is the third-tier to which all the district unions in a state are federated. The federation's board consists of elected chairman of the district unions and representatives of the State government. Its primary purpose is to maximize returns to the milk producer members through centralized marketing, purchase and quality control.

The Replication of Anand Pattern

Shri Lal Bahadur Sastri, the then Prime Minister of India, visited Anand in 1964 to inaugurate the cattle feed plant. Have been immensely impressed with the success of dairy co-operatives in Gujarat, he has advocated the replication of Anand-pattern dairy co-operatives throughout the country. Accordingly, the government of India established the National Dairy Development Board (NDDB) in 1965 to replicate this pattern throughout the country by implementing the OF programmes for this purpose. The Indian Dairy Corporation (IDC) was set up in 1970 to handle the donated commodities, the generation of funds and their disbursement for the dairy development programmes.

Table 1.6 displays growth of dairy cooperatives society. The number of dairy cooperatives was 63.4 thoU.S.A.nd initial year 1990-91 and 103.11 thoU.S.A.nd in 2001-02, registering 63 per cent increase. As regard to membership,

77.81 million in 1990-91 and 115.37 million in 2001-02 was regarding 48 per cent increase. Annual percentage growth rates both of cooperatives and membership are positive in the reference period; through in their magnitudes wide variations can be observed in the whole cooperative dairying in India is on growth trajectory.

Table: 1.6. Growth of Dairy Co-operatives and their Membership in India

Year	Societies (In 000's)	Yearly %age change	Members (In millions)	Yearly %age change
1990-91	63.4	....	77.81	....
1991-92	64.1	1.1	79.4	2.04
1992-93	65.5	2.18	83.7	5.41
1993-94	67.2	2.6	86.7	3.58
1994-95	69.8	3.87	89.09	2.75
1995-96	72.72	4.18	93.14	4.54
1996-97	77.99	7.24	96.05	3.12
1997-98	85.52	9.65	98.75	2.81
1998-99	87.76	2.62	131.4	33.06
1999-2000	101.42	15.56	129.08	1.77
2000-01	102.1	0.67	118.98	7.82
2001-02	103.31	1.18	115.37	3.03

Source: Indian Co-operative Movement- A profile, National Co-operative Union of India, Delhi, 1992-2004, pp 38-40.

Operation Flood Programmes

Operation Flood (OF), the world's largest dairy development programme ever undertaken, aims at setting up of a modern dairy industry to meet India's rapidly increasing need for milk and its products and making it viable and self sustaining growth. The project undertook the colossal task of upgrading and modernizing production, processing and marketing of milk with the assistance provided by the World Food

Programme (WFP).The aim was to create a "Flood" of rurally produced milk, assuring the farmer of remunerative price and ready market, and the urban consumer of wholesome milk at stable and reasonable prices by linking the main producing areas to the main consuming centers in urban areas.

Operation Flood-I (1970-81)

The programme laid emphasis on setting up of "Anand Pattern" rural milk producers' co-operative organizations to procure, process and market milk and to provide some of the essential technical input services for increasing milk production. OF-I was launched in 1970, following an agreement with the WFP, which undertook to provide 126,000 tonnes of Skim Milk Powder (SMP) and 42,000 tonnes of Butter Oil (BO) as aid for financing the programme. The programme involved organizing dairy co-operatives at the village level, providing the physical and institutional infrastructure for milk procurement, processing, marketing enhancing services at the union level and establishing of city dairies. The main thrust was to set up dairy co-operatives in the milk sheds, so as to link them to the four metro cities of Bombay, Calcutta, Delhi and Madaras, in which a commanding share of the milk markets was to be captured. The overall objective of Operation Flood-I was to lay the foundation of a modern dairy industry in India which would adequately meet the country's need for milk and milk products.

Funds for Operation Flood-I were generated by the sale of SMP and Butter Oil. A total of Rs. 116.54 crore was invested in the implementation of the programme. The achievements of OF-I are furnished in Table 1.7. By the end of OF-I about 13,300 DCS, 39 milk sheds were organized, enrolling 18 lakh farmer members. It achieved a peak milk procurement of 34 lakh litres per day (llpd) and marketing of 28 llpd.

Table 1.7. Dairy Development under Operation Flood Programmes in India

S. No.	Parameters	Phase-I Years		Phase-II Year	Phase-III Years					%ge increase in 2002 over 1971
		1971	1981	1985	1990	1994	1995	1996	2002	
		Panel A-Procurement								
1	2	3	4	5	6	7	8	9	10	11
2	Number of milk sheds	5	39	136	170	170	170	170	170	3400.00
3	Number of DCS (000's)	1.6	13.3	34.5	60.8	67.32	69.6	72.74	75.24	4700.25
4	Number of farmers membership (in lakhs)	2.8	17.5	36.3	70	86.9	90	93.14	95.24	3401.42
5	Average milk Procurement (llpd)	5.2	25.6	57.8	98.1	111.45	102	109.42	107	2057.69
6	Peak milk Procurement (1lpd)	6.5	34	79	120	130	116	140	151	2323.07
	Panel B-Procurement									
7	Rural dairies (llpd)	6.8	35.9	87.8	140.3	167.5	172	193.7	185.25	2724.26
8	Metro dairies (llpd)	10	29	35	37.9	38.3	52.3	72.4	86.5	765.00
	Panel C-Procurement									
9	Metro dairies (llpd)	N.A	21.8	29.5	30.6	32.34	35	38	45	206.42
10	Other cities & town (llpd)	0.9	6.1	20.6	41.9	53.9	59	61.38	52.26	5806.67
11	Total marketing	N.A	27.9	50.1	72.5	86.24	94	99.38	97.26	368.30
12	Milk dairy capacity (llpd)	N.A	261	507.5	663	831.5	842	74	956	366.30

1	2	3	4	5	6	7	8	9	10	11
13	Milk powder Production ('000 'tones/ year)	22.4	76.5	102	165	185	195	195.5	250	1016.07 %ge
	Panel D-Procurement									%ge increase in 2002 over 1981
14	Number of AI canters ('000,s	N.A	4.9	7.5	10.9	15.12	16.28	16.5	15.25	211.22
15	Number of AI Done (in lakhs)	N.A	4.98.2	13.3	30.1	....	37.9	39.5	41.5	733.33
16	Cattle feed capacity ('000' tonnes/ day)	N.A	1.7	3.3	4.3	4.7	4.9	5	5	294.11
17	Investment (Rs. Crores)	N.A	116.54	277.17	411.59	690.6	896.21	1303.1	151.6	130.08

Note: 1. llpd (lakh litres per day)
2. tpd (tonnes per day)

Source: Gupta, P.R., Operational Flood-Third phase, Dairy India Year Book, 5th Edition New Delhi, Rekha Printers, 1997, p.148, India Year Book, 2002 p. 47.

Operation Flood-II (1981-85)

The background of the institutional framework of OF-II essentially comprised of the successful replication of the Anand Pattern three-tier cooperative structure of societies, unions and federations. OF-II was designed to build on the foundation already laid by OF-I and the Indian Dairy Association (IDA) assisted dairy development projects in Karnataka, Rajasthan and Madhya Pradesh. The programme was approved by the Government of India, for implementation during the Sixth-plan period, with an outlay of Rs.273 crores. About US $ 150 millions were provided by the World Bank and the balance in the form of commodity assistance from the European Economic Community (EEC). The achievements under these programmes are given in Table 1.7.

OF-II helped to market milk in about 148 cities and towns with a total population of 15 million through a national milk grid, linking these towns and cities to 136 rural milk sheds. The project was extended crores 34500 village co-operative societies, covering 36 lakh farmer members. The peak milk procurement increased to a level of 79 llpd and milk marketing to 50 llpd.

Operation Flood-III (1985-2002)

The third phase aimed at consolidation of the gains of earlier twophases. The main focus of the programme was on achieving financial viability of the milk unions state federations and adopting the salient institutional characteristics of the "Anand Pattern" co-operatives. The OF-III programme was funded by a World Bank with a loan of US $ 365 millions, Rs.222.6 crores of Food-Aid (75,000 tonnes of milk powder and 75,000 tonnes of butter/butter oil) by the EEC and Rs. 207.7crores by NDDB from its own resources. The programme covered some 170 milk sheds of two countries by organizing 70,000 primary dairy cooperative societies.

The World Bank granted provisional extension of OF-III credit up to April 30, 1996. Its major emphasis was to consolidate the achievements gained during the earlier phases by improving the productivity and efficiency of the co-operative dairy sector and its institutional base for its long-term sustainability. Investments in OF-III were focused on strengthening the institutional management aspect of dairy co-operatives at various levels to establish financially strong, farmer owned and managed organization.

The OF-III also had provision for productivity enhancement inputs and institutional strengthening in the form of training, research, market promotion, monitoring and evaluation. Particular emphasis was placed on institutional and policy reforms. Efforts were made to expand infrastructural facilities in all major markets, linking them to milk sheds through the National Milk Grid (NMG) to ensure year-round stable milk supply. Marketing thus, becomes the linking force to improve procurement and strengthen the financial viability of the Unions. The role of NMG is crucial in ensuring the availability of milk to consumers and a remunerative price to milk producers by leveling out regional and seasonal imbalances in supply and demand. Marketing indigenous milk products forms an important part of the overall marketing strategy.

Table 1.7 displays progress of dairy development during three phases of OF under four broad parameters namely procurement, processing capacity, market and technical inputs. Between 1971 and 2000 all the indicators of milk procurement witnessed 4-digit increase in percentage terms. With regard processing capacity, the percentage growth in processing capacity was more in metro dairies (4-digit increased) than in rural dairies (3-digit increase). In milk marketing dairies in and towns either than those in metros had a cities better record (4-digit increased) and the same can be alert milk marketing powder. In permission of technical inputs there was a 3-digit increase in all the parameters.

Present Status of Operation Flood Programme

Flood Programme is now a movement of some 91.70 lakh rural families, who are the primary members of the milk cooperatives. As on march, 1995, about 69,600 DCS were organized into 170 milk sheds which procured 116 lakh kg of milk per day and marketed about 94 lakh liters per day (llpd) of liquid milk in over 600 cities and towns. Milk processing capacity of 842 tonnes per day was established. To operate the N.M.G. and balance regional and seasonal fluctuations in milk procurement and marketing, some 1108 road and rail milk tankers were provided for long distance transportation of liquid milk.

Adequate storage facilities were setup for store 33750 tonnes for milk powder and 4280 tonnes for butter to facilitate the operation of the NMG. With the increase in the production of milk and milk products, the country has been achieving a greater degree of self reliance than before. To stabilize the domestic prices of milk and milk products and exploit any export potential for Indian dairy products, the Government has recognized the NDDB as the canalizing agency for export.

For improving the productivity of dairy cattle and thereby milk production, the Flood Programmes provided animal health and breeding facilities. Nearly 40,313 DCS have been covered with the animal health programme, while 16280 DCS are provided with A.I. facilities. The bypass protein feed, developed by NDDB, has been increasingly adopted by farmers. This increases the protein conversion efficiency of the cattle feed by 33 per cent and dry fodder conversion by 30 per cent and minimizes the dry fodder requirement for milk production by 24 per cent. The treatment of straw with urea, a cheap and simple technique to raise the nutritional level of the straw, is being promoted. Feeding with urea-treated straws reduces the concentrate requirements by 33 per cent, minimizes wastage of straw and improves animal health. These technologies have implications on lowering the cost of milk production and thereby, maximizing returns to farmers.

Balanced cattle feed compounding capacity of 4905 tonnes per day has been setup and currently some 34576 DCS are engaged in marketing. Bypass protein technology has been introduced in 17 cattle feed plants, the feed being marketed thorough 5943 DCS. In addition, 8 Urea Molasses Block (UMB) plants with a total capacity of 72 tonnes per day have been established. Through Foot-and-Mouth Disease (FMD) control project, over 42 million vaccinations have been carried out. The Indian Immunological Technology Mission (IITM), Hyderabad, a subsidiary of NDDB, produced nearly 43.90 lakh doses of FMD vaccine, 8.92 lakh vials of veterinary formulations, 1,97,450 vials of rabies vaccine and 9.2 tonnes of vitamin premixes during the year 1995-96.

Dairy Development during Plan Periods

One of the indicators of a sector's importance is the budget allocation to that sector. The investment pattern in animal husbandry and dairying vis-à-vis total plan outlays during various plan periods is furnished in table: 1.8. The plan outlay of centrally sponsored schemes under animal husbandry and dairying increased from Rs. 22 crore, in the First Plan to Rs. 14.19 crore, and Rs. 1419 crore Eighth Plan. The outlay for dairying has increased from Rs. 7.81 crore in the First Plan to Rs. 1143 crore in the Eighth Plan. The allocation to animal husbandry and dairying as a percentage of total plan outlay varied from 1.48 per cent during the Fourth Plan to about 0.34 per cent during the Eighth Plan. Out lay and animal husbandry and dairying as percentages of total plan outlays were 0.11, 1.21, 1.05, 1.48, 1.10, 0.40, 0.25 and 0.37 in I, II, III, IV, V, VI, VII and VIII Plans respectively. It is district disturbing to note that there has been continuous decline in these percentages, indicating lower price investment in this sector. Although the dairy sector occupies a pivotal position and its contribution to the agricultural sector is the highest, the plan investments made so far does not appear commensurate with its contribution and future potential for growth and development.

The low productivity of Indian cattle has been the central concern of livestock policy throughout the last century. In the First Five-year Plan, the Key Village Scheme (KVS) was launched to improve breeding, feed and fodder availability, disease control, and milk production. To meet the need for milk of the urban areas, the government promoted state-owned dairy plants to handle milk procurement, processing, and marketing in 1959. Delhi Milk Scheme (DMS) was set up to supply milk to the urban population of Delhi, adoptions method of departmental milk procurement from the milk producing areas around Delhi by setting up its own milk collection and chilling centers. Though the collection was started from small milk vendors initially, it ultimately ended up creating big contractors who purchased milk from the small vendors and supplied it in bulk to the milk scheme. The same policies and strategies continued in the Second Five-year Plan. In 1976, the National Commission on Agriculture (NCA) concluded that the KVS could not meet its objectives due to a shortage of funds and hence it did not stress feed and fodder development and marketing of milk. The third plan emphasized the need to develop dual-purpose animals for milk as well as draft use. Crossbreeding of non-descript indigenous cattle was introduced during this plan. The Intensive Cattle Development Programme (ICDP) was launched in areas with high milk potential.

Table 1.8 presents plan outlays on animal husbandry and dairying as proportion of total plan outlays. It can be observed that the percentage share of this sector was 0.11 per cent in first plan which increased to the maximum of 1.47 per cent in fourth plan and than sided down to 0.25 per cent in annual plans 1990-1992 and slightly climbed upto 0.33 per cent in eighth plan. In sectoral allocation of public investment this sector was not received due attention.

Fodder Development

The programme also envisaged implementing Silivin pasture schemes for improvement of fodder availability. Silivipastures were developed on about 18,652 hectares of land. This programme was implemented both on community waste land and marginal lands of individual farmers in about 16700 and 16982 hectares respectively.

Dairy Development during the First Two Decades of Planning 1951-71

In India the government's attention to the dairy sector started right from the first Five Year Plan in 1951. The government decided to develope dairy sector through various schemes to increase milk production and to supply the dairy products to urban consumers at the lowest possible price. The major attempts are highlighted below.

Table 1.8. Plan-wise Outlay and Expenditure of Central Government and Its Sponsored Schemes under Animal Husbandry: From 1950 to 1997

(Rs. in crores)

Sl. No	Plan/Year	Total Plan Outlay	Animal Husbandry		Dairying		Total	Expenses (4+6)
			Outlay	Expenses	Outlay	Expenses	Outlay	
1	2	3	4	5	6	7	8 (3+5)	9 (4+6)
1	First Plan (1950-55)	19600	14.19	8.22	7.81	7.78	22.00 (0.11)	16.00
2	Second Plan (1955-60)	4600	38.50	21.42	17.44	12.05	55.94 (1.21)	33.47
3	Third Plan (1960-65)	8,576	54.44	43.40	36.08	33.60	90.52 (1.05)	77.00
4	Annual Plan (1966-67)	6,625	41.33	34.00	26.14	25.70	67.47 (1.01)	59.70
5	Fourth Plan (1967-72)	15,779	94.10	75.51	139.00	78.25	233.10 (1.47)	154.26
6	Fifth Plan	39,426	—	178.43	—	—	437.54 (1.10)	232.46
7	Sixth Plan (1980-85)	97,500	60.46	39.08	336.10	298.34	396.56 (0.40)	337.42
8	Seventh Plan (1985-90)	1,80,000	165.19	102.35	302.75	374.43	467.94 (0.25)	476.78
9	Annual Plan (1990-92)	—	101.68	79.46	177.16	119.42	278.84 (....)	198.88
10	Eighth Plan (1992-97)	422166	431.02	306.78	1143.89	818.18	1418.93 (0.33)	1124.98

Note: Figures in parentheses are total outlays on animal husbandry and dairying expressed as percentages of total plan outlays.

Source: *Basic Animal Husbandry Statistics 2004,* Department of Animal Husbandry and Dairying Ministry of Agriculture, GOVT. of India.

Technology Mission on Dairy Development

The OF programme prepared the ground for launching another massive Programme at the national level called

'Technology Mission on Dairy Development' (TMDD). The mission formally launched on 11th June, 1988 started functioning from June, 1989 with the main objective of accelerating the pace of growth of rural incomes and employment through dairy development. To achieve this end, the OF programmes are being dovetailed into other development programs such as the state programmes of animal husbandry and dairying, poverty alleviation programmes such as Integrated Rural Development Programme (IRDP), dairy research programmes, processing technology and product manufacturing programmes of the central research institutions, Agricultural Universities and National Dairy Development Board. The technology mission covered all the districts under OF programme in 1996-97. An additional 13,000 village level primary dairy co-operatives with organized by the State Governments in the OF districts. The mission achievements during the period 1988-89 to 2001-02 are shows in the Table 1.9.

The TMDD covered 242 districts in 1988 which steadily increased to 270 in 1996 and decreased to 265 in 2001. Correspondingly the number of milk sheds rose from 168 to 190 and decreased to 185 in the same periods. Another important achievement of technology mission was in the co-operative societies form about 49,077 to 70,000 and decreased to 65,000 in the same period. The average milk procurement was nearly doubled from 78 lakh liters per day to 150 llpd with the corresponding increase in the milk marketing from 67 llpd to120 llpd during 1988 to 2001. The processing capacity of rural and urban dairies was enlarged from 122 llpd and 37 llpd to 220 llpd and 48 llpd, and decreased 180 llpd and 40 llpd respectively. Thus when compared to metro dairies, the capacity of rural dairies increased at a faster rate. The average milk yield of both cows and buffaloes was doubled during the 10-years period, thanks to the measures initiated under technology mission.

Table 1.9. Role of Technology Mission on Dairy Development (TMDD): From 1988-2001

Sl. No	Components	1988	1995	1996	2001
1	Number of unions (milk sheds) covered	168	170 (1.19)	190 (11.76)	185 (-2.63)
2	Number of districts covered	242	266 (9.92)	270 (1.50)	265 (-1.85)
3	Number of co-operative societies in OF areas	49,077	72,744 (48.22)	70,000 (-3.77)	65,000 (-7.14)
4	Milk procurement (llpd)	78	109.40 (40.25)	150.00	152.00 (1.33)
	Average Peak	98	133.;6 (36.32)	200.00 (49.70)	210.00 (5.00)
5	Liquid milk marketing (llpd)	67	99.40 (48.35)	113.00 (13.68)	120.00 (6.19)
6	Processing capacity (llpd) Rural dairies	122	191.80 (57.21)	220 (14.70)	180 (-18.88)
	Metro dairies	37	72.90 (97.02)	48 (-34.15)	40 (-16.67)
7	Increase in average yields' (1991-92)	345	445 (29.00)	640 (43.82)	650 (1.56)
	Cows (lpa) Buffaloes (lpa)	692	811 (17.20)	1020 (25.77)	1250 (22.55)

Note: Figures in parentheses are percentage increases area the figures related to the increased immediately preceding year.
llpd = lakh liters per day;
lpa = liters per annum.

Source: Based on the report of the technical committee of direction for improvement of Animal Husbandry and Dairying statistics, 1994, Special Reports of NDDB on Operational flood- ii & iii and Dairy India, 1997, p.183; Dairy India Year book, 2002, p. 48.

Key Village Scheme

Key Village Scheme (KVS) was the most important component of the animal husbandry development programmes during the first three Five-year Plans. Initially, its main focus was on increasing the supply of breeding bulls in the country by setting up bull breeding farms in the major cattle tracts. Gradually the KVS was transformed into a more comprehensive programme for general cattle development intended to improve the productivity of cattle by giving simultaneous attention to better feeding, improved breeding, effective disease control measures, scientific management practices and organized marketing facilities. Towards the end of the second plan nearly 600 KVS centres were functioning in the country covering an estimated 6 million cows and she-buffaloes, which was about 10 per cent of the total stock. However, in the absence of stable and remunerative market for milk, production remained more or less stagnant. During the two decades between 1951 and 1970 milk production grew by barely 1 per cent annually while per capita milk availability declined.

Intensive Cattle Development Project

The perceived failure of the KVS to make significant impact and the shortage of milk in the rapidly growing urban areas led to the formulation of the Intensive Cattle Development Project (ICDP). Whose primary purpose was to increase the production of milk to feed public sector dairy plants in the hinterlands of the main urban centers? Consequently they placed great emphasis on cross-breeding in indigenous cows with exotic dairy breeds and tended to be concentrated in milk shed areas of large cities and towns. Animal health and breed improvement remained common elements in the project. The scheme envisaged provision of all necessary inputs and services simultaneously to milk producers. The ICDP is, thus, distinguished from KVS by the shift of attention to the cross-breeding component and closer linkup with dairying and urban milk supply programmes. In areas where ICDP existed, the KVS was merged into the farmer. In areas where ICDP did not exist, KVS was continued in the original form.

Establishment of Cattle Colonies and Milk Schemes

During 1960s various state governments tried out different strategies to develop dairying, which include establishing dairies run by their own departments, setting-up cattle colonies in Bombay, Calcutta and Madras. These government projects had extreme difficulties in organizing rural milk procurement and running milk schemes economically. Yet none concentrated on creating and organized system for procurement of milk which was left to contractors and middlemen. Perishable nature of milk and its scarcity gave the milk vendors leverage which they used to a considerable advantage. This left government run dairy plants to use large quantities of relatively cheap, commercially imported milk powder, which resulted in a decline in domestic milk production. All these factors combinely left Indian dairying in most unsatisfactory low level equilibrium.

Dairying in Post-Reform Period

The third phase of Indian dairy development started in 1991, when the Government of India introduced major trade policy reforms farming increasing privatization and liberalization of the economy. The dairy industry was delicenced in 1991 with a view to encourage private sector participation and investment in the sector. Two major events were turning points in the post-reform period in dairy sector.

Introduction of Milk and Milk Products Order

The government introduced the Milk and Milk Products Order (MMPO) in 1992 under the essential commodities Act, 1995 to regulate the production of milk and dairy products. The order required permission from State/Central registration authorities to set up units handling more than 10,000 litres of milk per day or solids up to 500 tonnes per annum, depending on the capacity of the plant. The order included sanitary and hygienic regulations to ensure product quality.

However, concerns were raised about these government controls and licensing requirements for restricting large

Indian and multinational firms from making significant investments in this sector. The government, therefore, amended the MMPO in March, 2002 and restrictions on setting up milk processing and milk product manufacturing plants were removed and the concept of milk shed was abolished. This amendment is expected to facilitate the entry of large companies which would definitely increase competition in the domestic markets.

Uruguay Round Agreement on Agriculture

The second major development in Indian dairy sector policy came when India signed the Uruguay Round Agreement on Agriculture (URAA) in 1994 and became a member of the World Trade Organization (WTO), which made India open up its dairy sector to world markets. The import and export of dairy products were delicenced and trade in dairy products was allowed freely with certain inspection requirements. The first major step was taken in 1994-95, when the import of skim milk powder and butter oil was decanlized. Restrictions on the remaining products were removed in April 2002. Now India has bound its import tariffs for dairy products at low levels according to the Uruguay round decisions.

Perspective 2010

Operation Flood (1970-1996) paved the way to take up new initiatives and create new conditions to firm-up India's world leadership in milk production. The new challenge for the dairy industry was to explore ways to emerge stronger using the network created under OF. The response is 'Perspective 2010', a plan that attempts to take the dairy co-operative movement to its highest potential. 'Perspective 2010' focuses on four key areas. These include strengthening co-operative business, production enhancement, assuring quality and creating a national information network. NDDB facilitated the planning process sand will provide technical support and need-based finance for implementing 'Perspective2010'.

Strengthening Co-operative Business

Perspective 2010 aims to recruit, train and motivate increasing number of women to work for co-operatives, to achieve significant improvement in dairy husbandry as they primarily shoulder animal husbandry related responsibilities in rural India. It visualizes the consolidation and growth in milk and milk-product marketing, promoting better equity for regional co-operative brand and developing qualified and skilled manpower. It also aims at persuading the state and central governments to remove the shackles on co-operative laws so that co-operatives can compete on equal terms with other forms of enterprise. Expanding the market is a major target of perspective 2010 which offers financial and technical help to milk unions and federations in areas such as sales promotion, consumer education, infrastructure development, etc. As part of sales promotion it recommends standardization of artwork, colour, logo and retail outlet design across regional co-operative brands with a view to promote better recall by consumers under a common mnemonic umbrella. Another target is increasing women membership in dairy co-operatives to 50 per cent and improving women participation in the governance of dairy co-operatives at all levels.

Production Enhancement

Perspective 2010 stresses to improve the production potential of indigenous breeds of cattle such as Sahiwal, gir, Rathi and Kankrej and breeds of buffalo such a Murrah, Mehsana and Jaffarbadi through appropriate selection programme. It gives proper direction to crossbreeding technology to increase production in such a way that crossing of non-descript cattle with Holestein Friesian in areas with adequate feed and fodder and with Jersey in resource-poor areas. As a step to increase production and availability of fodder, it appeals to unions, NGOs and co-operatives to put common property area under improved pasture and fodder tree. It promotes first aid, coverage through village level societies and diseases free zones in the country.

Quality Assurance Programmes

As part of increasing quality it facilitates improvement of hygiene, sanitation, food safety and operating efficiency in the dairy plants and sensitizes dairy personal to product quality aspects as per international standards. It promotes encouragement of quality incentives supported by educational programmes for dairy co-operative society staff, transporters and farmer producers. Quality is assured through facilitation dairy co-operative in ISO 9000-2000 (Quality Management Systems), ISO HACCP Safety Management Systems Certification which can maintain the required plant conditions under the accreditation on a sustainable basis.

Information and Development Research

'Perspective 2010' plans to link large co-operatives, unions, federations and NDDB in a national network that collects and disseminates information to all. It ensures the availability of analytical information for policy planning and implementation. The integrated dairy industry information service facilities decision-making at various levels in co-operative institutions with the help of an extensive on-line computer network that analyses relevant data obtained from DCS, District Milk Producers' Union (DMPU), State Milk Marketing Federations (SMMF), NDDB and Research institutions. Perspective 2010 also proposes the need of a national database that generates data on milk supply (producer, animal and village data), data on milk and milk product demand (consumer and urban data), performance data (societies, unions and federations), and secondary data (domestic and international).

The Challenges Ahead in Dairy Sector in the Country

They are:

(*i*) The Indian dairy sector comprises mainly of farmers with small landholding and low productivity milch animals.

(*ii*) There are many milk producers clustered within villages, away from the urban consumption centres. The real challenge is to organize and weave a network making the system work as an industry.

(*iii*) Every small milk producer is dependent upon the dairy income from default in payment foils his family budget.

(*iv*) Milch animal, an asset of the dairy farmer, is under continuous

(*v*) Threat from local and infectious diseases causing mortality and morbidity poor farmers are unable to afford such calamities.

(*vi*) Lack of infrastructure facilities hinders the growth of the rural dairy farmers.

(*vii*) In rural milk markets, there are many producers but only a few buyers, resulting in distress sales due to lack of proper marketing facility; and

(*viii*) In the ERA of globalization and liberalization, those who produce quality items at cheaper cost gain market world-wide access. The developing countries, though produce milk cost effectively, suffer from non-tariff barriers in international trade under Sanitary and Phytosanitary Measures Agreement (SPMA). The stringent food safety standards are beyond the reach of Indian dairy industry.

Responses to Meet the Challenges

The following measures, if taken up, to put the industry on the sight path.

Enhancement of Productivity through Breed Improvement

Cross-breeding of low-yield indigenous breeds with high-yield exotic breeds has been widely acknowledged as a valuable strategy to improve animal productivity. The focus of cross breeding research is mainly on cattle because of their dual role of milk production and use as draught animals in the crop sector. A number of cross breeds evolved with improved predestining potential include Haryana × Friesian, Haryana × Brown Swiss, Rathi × Jersey, Gir × Jersey, Gir × Friesian and Sahiwal × Jersey.

High Rate of Productivity through Improved Feeding Practices

Cattle are to large gap the problem of underfeeding of between requirement and availability of feed and fodder in the country. This problem of underfeeding can partly be overcome through technological interventions such as biological and chemical treatment of feed and fodder. For example, urea molasses treatment can improve the quality of dry fodders and straws of wheat and paddy. Urea molasses mineral lick can be used as a supplement to straws, which can help improve the digestion of cellulose fodders and the efficiency of rumen micro-flora. Urea treatment it is reported that, reduces green fodder requirement by about 20-40 per cent and increases yield by 10-20 per cent. Further by-pass protein technology reduces concentrate requirement 40 per cent and dry matter requirement 24 per cent. In addition, it is possible to increase the productivity of crops like sugarcane, sun hemp, cowpea, carrot, cauliflower and turnip.

Increase of Productivity through Improved Health Services

Unsanitary conditions block the full realization of genetic potential of milch animals. Proper health care management of the animals is possible through a three-tier treatment system. First-tier is the aid at the village level through a local resident trained as veterinary assistant to diagnose and provide immediate relief to sick animals. Mobile veterinary service, the second-tier, should visit villages on scheduled days and time. Three-tier is the vaccination against the epidemics.

Value Addition to Farmer's Price Realization

In the globalization ERA, value addition is possible by marketing of branded milk products. Branded western and ethnic dairy products are witnessing rising demand and increased acceptance, especially by urban consumers. Ethnic products viz. sweets, *paneer, dahi* (yogurt), etc. offer growing opportunity for the organized sector. The success of branded dahi, flavored milk and traditional sweets such as gulabjam, doodhpeda and burfi suggest the potential for introducing such products to masses. A total quality management approach to improve the quality of milk products has to follow the entire chain from milk production to the consumption point.

Dairy Development through the Provision of Better Infrastructure Facilities

Transport road infrastructure reduces transport time, enabling the produces and market functionaries to realize better price for their products etc. In the dairy sector it is found that wireless communication can trace milk carriers for breakdown, enable the veterinarians to render emergency service to sick animals in remote rural areas.

Application of Information Technology at Rural Level

Information technology helped dairy co-operatives to change their rural face. It facilitates computerized milk collection units to provide information about fat content of milk and volume and to calculate of amount payable on the spot. The process has increased the trust, transparency and efficiency of milk collection. Introduction of internet would link all e-transactions between village co-operatives and processing units.

Production and Marketing of Milk through the Formation of Dairy Co-operatives

In India milk production is mainly routed through co-operative system. Which is the basis of OF, the co-operatives assured the marketing of milk from the small dairy farmers, offering them a reasonable price. They also provide various inputs to dairy farmers at affordable prices. They promote cooperative spirit essential for their survival

Initiatives at the Government Level

To increase the efficiency of livestock sector and to make it competitive in the world market, the Government of India has created the Department of Animal Husbandry and Dairying (DAHD) in the Ministry of Agricultures in 1991. It primarily supplements and complements the efforts of state governments in enhancing productivity levels of livestock through genetic upgrading, increasing the availability of feed and fodder, providing veterinary care, processing facilities, strengthening marketing infrastructure, improving the database for livestock products and so on.

While formulating the Tenth Five-year Plan, the Government of India stressed the following strategies;

- Conservation of native livestock to maintain diversity of breeds;
- Immunization programme against diseases and creation of disease free zones;
- Enhancement of feed/fodder production and improvement of common property resources;
- Creation of National Animal Health and Production Information System; and
- A transition from subsistent livestock farming to sustainable and viable livestock farming.

The All India Rural Credit Review Committee (AIRCRC) also emphasized the need for providing subsidiary occupation like dairy farming to the peasants. Agriculture creates problems of unemployment and underemployment, seasonal employment and disguised unemployment to rural people constituting 70 per cent of total population. Young people from rural areas migrate to towns or cities for work as rural economy is in shambles due to the vagaries of climate. Dairy enterprise is a solution to overcome such problems and besides being an effective tool to improve socioeconomic conditions of farmers.

Milk Consumption Pattern in India *vis-à-vis* other Countries

On an average an Indian requires 280 grams of milk daily for his balanced diet, as against which average consumption of milk is 121 grams per day, leaving a wider gap. Punjab tops the list in the consumption of milk and milk products with a daily consumption of 220 grams followed by Gujarat and Maharashtra with 140 and 100 grams respectively. Kerala and Bihar are far behind with less than 20 grams. The consumption of milk in foreign countries is much higher at 714 grams in Switzerland, 637 grams in New Zealand, 623 grams in United States and 509 grams in United Kingdom. Through per capita consumption of milk and milk products in India is the highest in Asia. Yet, it is still below the world average of 285 grams per day and the minimum nutritional requirement of 280 grams per day as recommended by the Indian Council of Market and Research (ICMR) 2005.

Prospects of Dairy Industry in India

In 1991 as part of the economic reforms, the dairy sector was delicensed, opening up the industry to private entrepreneurs including multinationals which are the foreign companies were allowed to raise their equity holdings up to 51 per cent. The basic philosophy underlying delicensing was to encourage the competition in procurement of milk, thus, enhancing value for both producers and consumers. It was also expected to spur increasing inflow of capital and new technologies. Delicensing did have the intended effect of attracting private sector investments into the dairy industry. Within a year, over 100 new dairy processing plants were established in different parts of the country, most of which were designed to manufacture a range of high value-added products. As private enterprises began operating, the new entrants in some areas were accused of "poaching" into co-operative territory. Moreover, concerns about excessive capacity for milk products and private traders' misconduct (sales of adulterated/contaminated milk) prompted need for some control over the market. Consequently, the government of India promulgated the MMPO in June 1992.

MMPO prescribes certain provisions for the orderly development of the dairy industry. Its key feature is that all plants handling between 10,000 and 75,000 liters per day or producing between 5000 and 3,750 tonnes of milk solids per year are required to be registered with State authorities, while those processing over 75,000 litres per day or producing more than 3,750 tonnes per year of milk solids require registration with the central government. The government should create a level playing field where dairy co-operatives or for that matter, agriculture co-operatives can effectively compete in the liberalized environment with private dairies and multinational food processing companies. This can be achieved by adopting a progressive co-operative law both at the centre and the state levels, enabling co-operatives to operate as autonomous bodies. Equally important, government was urged to shift milk powder imports to the negative list and impose a minimum basic import duty of 40 per cent advalorem, which will ensure that subsidized exports of milk products from developed dairying nations do not adversely affect India's quest for self sufficiency in milk production.

It is expected that the proposed measures could help in defining the role of the Indian union and its federal states on some of the most significant issues that concern dairying. This would lead us to a achieve the common goal of higher milk production and productivity, greater efficiency in milk handling, processing and marketing and an increased availability of liquid milk of desired quality to the general public at affordable prices.

PART-B : DAIRY DEVELOPMENT IN ANDHRA PRADESH

Significance of Dairying in Andhra Pradesh

The significance of dairying Andhra Pradesh needs no emphasis in the given structure of economy and the distribution of population and labour force. Abhorring pressure on land and consequent reduction in the size of land-holding are compelling reasons for farmers to take up dairying as an occupation. Uncertainties in rainfall and inadequacy of irrigation facilities in the State made animal husbandry and dairying a subsidiary to agriculture, promoting the rural employment and income. A.P. is the fifth largest state in India in respect of area and population. The coastal belt of A.P., known as Andhra region with its fertile land and substantial stable irrigation facilities, provides the necessary infrastructure for dairying whereas the other two regions, Telangana and Rayalaseema, lag behind in these respects. Though Telangana is endowed with vast forest area and cultivable land, the poor irrigational facilities have hindered the dairy development. The situation is worse in Rayalaseema due to unfavorable ecological setting.

A.P. State Domestic Product and Its Sectoral Shares

Agriculture is the mainstay in A.P. in terms of its contribution to states income and also employment, as evidenced by data domestic product and employment shown in Table 1.10.

Though the share of agriculture, over the years, in state domestic product has fallen to about 32 per cent, its contribution to employment it still ranks high with 70 per cent.

Table 1.11 shows growth of population, milk production and per capita availability of milk since 1951. From the perU.S.A.l of the table it is clear that the growth rate of per capita available of milk is lower than the growth rate of population, 108 gms/day was lowest in 1971 which increased to 147 gms/day in 2001. Milk production witnessed a growth rate of 37.50 Per cent between 1971 and 1981 and 30.33 per cent and between 1981 and 1991, 28 Per cent between 1999 and 2001. The growth of per capita milk availability was only 11.56 per cent, 7.32 per cent and 13.20 per cent respectively during the corresponding periods. Moreover, milk production slackened in 1980s, pushing dairy growth rates of milk production and per capita milk availability. Hence,

Table 1.10. Sectoral Distribution of the State Domestic Product and Employment in Andhra Pradesh in Select Years

(in per cent)

Sl. No.	Sectors	State domestic product			Employment		
		1950-1951	1990-1991	1999-2000	1950-1951	1990-1991	1999-2000
1.	Primary sector (including Agriculture and allied activities)	52.00	38.24	32.22	70.72	71.25	68.93
2.	Secondary sector (including manufacturing)	15.00	19.38	22.26	11.16	10.48	9.72
3.	Tertiary sector (including services)	33.00	42.38	45.52	18.12	18.27	21.35
	Total	100.00	100.00	100.00	100.00	100.00	100.00

Source: *Statistical Abstract of Andhra Pradesh,* Directorate of Economics and Statistics, Government of Andhra Pradesh, Hyderabad, 1995,2000.

the need to boost milk production so as to increase per daily milk availability.

GROWTH AND STRUCTURE OF DAIRYING A.P.

Prior to the entry of the state government into the dairy business of dairying, it was mainly in the hands of unorganized individual entrepreneurs with the result of quality of milk was pool were high. The state government organized the dairy industry by setting up a department in the government and also by encouraging the establishment of village level milk producer's cooperative societies for the development of dairy industry on democratic lines. Before 1960, dairy development was one of the many subjects in the state attached to the Department of Animal Husbandry. Initially, a pilot milk supply scheme was started during 1960-61 in Hyderabad as a prelude to the implementation of the integrated milk project in 1964. The Department of Animal Husbandry was entrusted with the responsibility of implementing the pilot milk supply scheme. In 1964, a milk powder factory at Vijayawada and a central dairy at Hyderabad were setup. Simultaneously, the Milk Powder Factory, Vijayawada and the chilling centre in Krishna district were also taken up. The work in respect of the cooperative dairies at Nellore, Chittor, Warangal and Kurnool was completed.

Table 1.11. Growth of Population, Milk Production and Per Capita Availability in Andhra Pradesh in Select Years

Sl. No	Item	1951	1961	1971	1981	1991	2001
1.	Total population	311.15	359.83	435.03	535.50	665.08	768
2.	Growth rate of population (per cent)	14.02	15.65	20.90	23.10	24.20	15.47
3.	Total milk production (in million tones)	NA	NA	1.68	2.31	3.01	3.855
4.	Growth rate of milk production (percent)	—	—	—	37.5	30.30	28.07
5.	Per capita milk availability (gm/day)	NA	NA	108.45	120.99	129.85	147
6.	Growth of per capita availability of milk (per cent)	—	—	—	11.56	7.32	13.20

Source: 1. *Statistical Abstract of Andhra Pradesh,* Directorate of Economics and Statistics, Government of Andhra Pradesh, Hyderabad, 1995, 2002.
2. *Dairying in India*, XXVIth Dairy Industry Conference, New Delhi, 1995, pp. 3.7-3.13
3. Dairy India year book 2002, p. 12.

Formation of the Andhra Pradesh Dairy Development Corporation (APDDC)

During the years 1972-73, the activities of the dairy development expanded enormously and its transactions rose to the tune of Rs.13 crore annually. Therefore, the government felt that an organization of such size could no longer be run under a Government Department and that it should function on its own as a corporation on commercial lines. The corporate setup would have an added advantage of availing itself of institutional finances for rapid development of the dairy industry in the state. In the Indian Dairy Corporation (IDC), the UNICEF and Central Government. The APDDC came into existence on February7, 1974, with an authorized share capital of Rs. 5 crore which was subsequently increased to Rs.15 crore with 15 lakh of equity shares of Rs. 100 each to be fully subscribed by the State Government.

Objectives and Role of APDDC

The APDDC was established with the objective of materializing the socio-economic change with massive dairy development programmes and providing employment and income to the producers throughout the year, besides improving the nutritional standards of people by supplying required quantity of milk to the cities and towns. Precisely "Changing life styles of people" was the motto of APDDC. Collecting milk at the door steps of producer in remote widely scattered Villages, the corporation has been supplying milk daily to urban consumers.

Formation of Dairy Development Co-operative Federation (APDDCF)

The State Government decided to convert the APDDC into the dairy Development Co-operative Federation Limited (APDDCF) on may 5th 1981. The APDDCF had taken over the business of APDDC with effect from October, 1981. The established of APDDCF is in conformity with objectives laid down by the central government. Moreover, establishment of such federation was also a pre-requisite to get funds from the IDC. It is the policy of the IDC to sanction funds only to the co-operative Federation.

Objectives of the Federation

Its objectives are:

(*i*) To carry out activities for promoting production, procurement, processing and marketing of milk and by-products for the economic development of the farming community;

(*ii*) To purchase commodities from members and others and pool, process, manufacture and distribute the same;

(*iiii*) To study the problems of mutual interest relating to production, procurement and the marketing of dairy products;

(*iv*) To provide veterinary aid and artificial insemination and undertake animal husbandry activities, so as to improve the animal health care and disease control activities;

(*v*) To make arrangements to impart training to staff of all the member unions and societies;

(*vi*) To assist the member unions in respect of promoting the organization of primary societies, plan the development strategy, render technical, administrative, financial and other assistance and advice unions on price fixation, public relations and allied matters; and

(*vii*) To undertake periodical supervision of the member unions and other societies affiliated to them and recommend the liquidation of any union or primary co-operative society.

Membership of Federation

The membership of the federation is open to three types of bodies as mentioned hereunder:

(*a*) The State Government of Andhra Pradesh;

(*b*) The District Co-operative Milk Producers Union, procuring lakh litres of milk per annum and having business dealings with the federation for a period of two years; and (*c*) The District Co-operative milk Unions not falling under the above category admitted as nominal members.

Each participating member society should hold at least twenty shares of a value of Rs.1000 each. While the Guntur union was the only are established under OF-I, 16 Districts were covered under Integrated Dairy Projects (IDP) assisted by National Co-operative Dairy Development Corporation (NCDDC) and Government of AP. The milk was collected from about 6 to 20 lakhs milk producers through 8482 collection centers organized under 398 milk sheds covering 112,375 villages. There are about 345 All Women Dairy Co-operative Societies (AWDCS) of which 109 are in chittoor district. There are 76,300 rural women members in all the milk collection centers. APDDCF has 10 district unions covering 14 districts, 3 of milk shed districts and 7 non- milk-shed districts. There are two major dairies, 10 district unions and one mini dairy with a total installed capacity of 23.87 lakh lpd supported by 72 milk centers. The Milk products factory produces milk products like skimpowder, whole milk powder, baby food icecream, butter, ghee, kulfi, buttermilk, khova, cheese. Considering the significance of premixed cattle feed the Federation has established 8 cattle feed mixing plants with capacity of metric tonnes per day. To encourage the farmers to raise high yielding fodder. Fodder seed production farm has been established at Nakrekal in Guntur district. There is also a buffalo breeding center at Nakrekal. At the village level 520 AI centers and 2385 first aid centers are functioning. There are training centers of which 4 are located at Visakhapatnam, Guntur, Mydukur, and Bhangir and are operating under the Federation. Besides, Krishna union has its own training center at Vijayawada and the Chittoor union has established a separate training facility.

The Federation sells 2398 lakh liters of milk annually through various distribution outlets located in 102 towns and cities covering 4.5 lakh consumer families. The turnover of the Federation/Union is Rs. 380 crores per annum. APDDCF received financial assistance from NDDB to the tune of Rs. 50 crores for OF programmes as 42 crores for programmes OF-II and Rs. 42.86 crores for OF-III programmes Rs. 37.54 crores. The Federation received Rs. 394 lakhs from government under various schemes during 1991-92 to 1994-95.

Dairy Development during Plan Periods in A.P.

The approach of the government was to develop dairying as an industry which will look after the rural problems like poverty and unemployment, to meet growing demand for milk especially in urban areas, to develop the industry as an expert industry and rubbing greater employment diversification through dairying in rural areas.

Second Five-year Plan (1956-61)

During the second Five year plan emphasis was given to the problem of supply of pure milk at reasonable rates. A sum of Rs. 80.15 lakhs was provided for the implementation of pure seven schemes designed to improve the supply of pure milk at reasonable rates to both milk producer and the consumer. But during the course of the plan three schemes involving Rs. 13.40 lakh were either dropped or deferred, and for these schemes the allocation was reduced by about Rs. 19.37 lakhs on the basis of their performance and requirements. On the other hand, four new schemes costing Rs. 744 lakhs were included.

Third Five-year Plan (1961-66)

The third five-year Plan gave emphasis for the milk supply and the production of integrated services of inputs. The Plan assumed that demand for livestock World grow at 5.5 to 6.4 per cent per annum. An amount of Rs. 255.98 lakhs was spent during this plan to meet the milk requirements of the twin cities i.e., Hyderabad and Secundrabad. Integrated milk project costing Rs. 4.35 cores was taken up in the Plan to procure milk in the Vijayawada milk-shed area and transport into the twin cities by rail and road. Besides the integrated milk project, a few schemes were also taken up to meet the milk needs of other urban areas.

Fourth Five-year Plan (1969-74)

An investment of Rs. 385.46 lakhs was provided out of which the Integrated Milk Projects (IMP) at Hyderabad and Vijayawada accounted for Rs. 222.07 lakhs. The milk supply to the twin cities and the milk supply factory at Vijayawada accounted for one crore each. An amount of Rs. 163.39 lakhs was earmarked to Integrated Milk Supply Scheme (IMSS) to establish a number of milk chilling centres. Besides IMP few projects were also taken up to meet the milk needs of other urban areas. The Fourth Plan followed regional approach for the development of dairying:

(*i*) The first level being meant primarily for local consumption and activities are highly localized to a few urban small centers; and

(*ii*) The second level being creation of the bigger network to provide milk for metropolitan cities and bigger urban centers like Hyderabad and Visakhapatnam.

Fifth Five-year Plan (1974-78)

The Fifth Plan aimed at alleviating poverty, particularly in rural areas, and identified dairy development to be a major important instrument for this, differentiating between the Programmes of Animal Husbandry (AH) and those of dairy development. The programmes in this plan focused on establishment of chilling centers and milk powder facilities. During the Plan period it was expected that the installed capacity of all the dairying units in the state would increase from 10.25 lakh liters per day at the end of the Fifth Plan to 19.00 lakhs liters per day at the end of 1982-83.

Sixth Five-year Plan (1980-85)

The objective of sixth plan with regard dairying was "to develop and strength the achievements of previous years so that all surplus milk produced in the rural areas is put to proper use". The approach of the Sixth plan was to provide more employment opportunity by way of subsidiary employment to small and marginal farmers and to increase the milk supply to the urban population. The expenditure on dairying during sixth plan was Rs. 1418 lakhs.

Seventh Five-year Plan (1985-90)

The Plan stated that dairying is the main source of supplementary income in the rural areas especially women who to the extent of about 80 per cent work in dairying and majority belong to marginal farmers. It was planned to extend milk procurement to about 1560 villages benefiting 1.5 lakhs milk producer families. Consequently the dairy infrastructures were expanded and strengthen.

Annual Plans (1990-92)

During the plan an amount of Rs. 197 lakhs was spent for different activities of dairy development resulting in increase in procurement four about 28.21 crore liter by the end of the Seventh Plan to 29.22 crore liters by 1992.

Eighth Five-year Plan (1992-97)

The Eighth Five-year Plan aimed at "accelerate the pace of increasing rural income and employment through dairying by improving productivity". For achieving this plan stated that the dairy programmes will be dovetailed with the poverty alleviation programmes. The plan also aimed at consolidating the cooperative dairy sector for ensuring its growth on a self-sustaining basis. An amount of Rs.1000 lakhs was allotted to dairying which constituted 0.10 per cent of the total plan outlay. The plan envisaged different programmes like extension of technical inputs, NCDC (National Co-operative Development Council) assisted IDP for non-OF district, processing infrastructure, modernization and expansion of Hyderabad dairy.

Dairy Development Under OF

The OF project was ushered in the State to strengthen the dairy industry and stimulate the growth of production of milk and milk products. OF-I was implemented in later 1970s (1975-76) mainly to eliminate the existing constraints to both villages and district level co-operatives. Under this programme a two-tier cooperative dairy structure Anand pattern established first time in Guntur milk shed, Guntur district. The Government of AP entrusted the implementation responsibility of OF to the Andhra Pradesh Dairy development Corporation.

Operation Flood-I

Started in the year 1975, the total investment under OF-I in A P was Rs. 474 lakhs at the end of January, 1981. Under OF-I project a feeder balancing dairy of 1.5 lakh liters per day was established at Sangam, Vadlamudi in Guntur District. Thus, OF-I project, paved the way for the eliminating sales and marketing bottlenecks. OF-I, on one hand, induced the farmers to invest in dairy development by providing financial assistance and, on the other hand established milk product facility. Mini chilling centers were established in various towns of costal Andhara such as Kuppam, Senthampet, Kothapalli, Arkad, Rampa Chodavaram and Khannapuram and in various Telangana towns such as Turunagam, Bhadrachalam and Adilabad. Monthly card system was introduced in the cities in 1975, in order to ensure an assured supply of milk through 299 sale points. Free pack machines were also introduced for supplying milk to the customers.

Operation Flood-II

Huge expansion investment, APDDCF subsequently, the milk product factories at the district level were registered as co-operative unions on Anand pattern for implementation of dairy development programmes.

In October, 1978, Government of India sanctioned an outlay of Rs. 4855 millions, funded by the IDC was the project authority and responsible for implementing the project in the country in cooperation with the State Governments. 16

districts were selected in the State in 1981, for implementation of the programme. To implement the OF-II Programme the APDDCF was registered on 5-5-1981. The salient features of the OF-II project were.

- Increasing milk by 10.74 lakh liters per day;
- Establishment of 5040 "Anand Pattern" dairy co-operative societies including reorganization of existing milk collection units;
- Establishments/expansion of capacity of chilling centers.
- Construction/expansion of an incremental 300 metric tonnes capacity of balanced cattle feed per day;
- Coverage of 12.10 lakh milch animals under technical inputs programme;
- Contribution of 12 lakh liters of milk per day through rail tankers and other one lakh liters per day through road milk tankers;
- Training of spearhead teams DCS; and
- Provision of technical assistance and implementation.

OF-II programme covered 16 selected districts grouped into 9 milk sheds, 19,094 in habituated villages and 23 urban centres.

The total outlay OF-II project in the state was Rs. 785.08 lakh. Marched from assistance India Dairy Corporation with 70 and 30 per cent of loan and grant components respectively. The loan carried an interest of 8.5 per cent per annum and the principal is repayable in 30 half-yearly installments after a moratorium of 5 years.

Operation Flood III

Nellore district not included in OF-III project area, was included for coverage under OF-II programme. It was proposed to organize 5030 milk Producer Cooperatives with a total membership of 6.71 lakh by 1992, and to cover 1330 milk producer cooperative societies under Artificial Insemina-

Table 1.12. District Co-operative Milk Producers Union (Anand Pattern Dairy Unions) in A P Covered under OF-II, 2001-02

Sl. No.	Name of the union	District covered	OF- phase covered	No. of districts
1	Visakha Union	Srikakulam, Vijayanagaram and Visakha	II	3
2	Godavari Union	East Godavari &West Godavari Distt.	II	2
3	Krishna Union	Krishna Distt.	II	1
4	Guntur Union	Guntur Distt.	II	1
5	Prakasam Union	Prakasam Distt.	II	1
6	Nellore Union	Nellore Distt.	II	1
7	Chittoor Union	Chittor Distt.	II	1
8	Kadapa Union	Kadapa Distt.	II	1
9	Kurnool Union	Kurnool Distt.	II	1
10	Nalgonda & Ranga Reddy Unions	Nalgonda and Ranga Reddy Distt.	II	2
	Total	14	...	14

Source: Compiled from the records of APDDCF, Hyderabad.

tion (AI) with a total project cost of Rs.75.51 crores. The salient features of OF-III were:

- Expansion of liquid milk plant at Visakahapatnam to handle 2.0 lakh litres per day;
- Establishment of the milk products factory in Ongole, and Prakasam districts;
- Establishment of chilling center at Porumamilla in kadapa district and another in Prakasam district;
- Strengthening of milk chilling centre at Narasapatnam;

- Strengthening of liquid milk processing, packing and distribution at Kakinada, Bhimavaram and Badvel;
- Strengthening and expansion of spray capacity of milk product factory at Chittoor; and
- Expansion of milk products factory at Vijayawada with the estimate project cost of Rs. 75.51 crores.

Development of Dairy Infrastructure in the State

The infrastructure for dairy development consists of dairy units, milk product factories, and feed machine and livestock development centers. Table 1.13 shows the infrastructural facilities for dairy development in A.P. The number of dairy units increased from 44 to 76 between 1975-76 and 1980-81 declined to 54 in 1985-86, soared up to 70 by 1995-96, declined to 65 in 1996-97, and still for the further to 33 in 2000-01. The intermittent decline was due to closure of uneconomical units. But the installed steadily increased from 6.92 lakh litters per day in 1975-76 to 22.97lakh litres per

Table 1.13. Development of Dairy Infrastructure in A.P. in Select Years

Sl. No	Particulars	1975-76	1980-81	1985-86	1990-91	1995-96	1996-97	2000-01
1.	No. of Dairy Units	44	76 (72.72)	54 (-29.95)	56 (3.70)	70 (25)	65 (-7.86)	33 (-49.24)
2.	No. of Major Dairies	—	1 (0)	1 (0)	2 (100)	2 (0)	2 (0)	2 (0)
3.	Installed Capacity (lakh liters per day)	6.92	13.53 (95.52)	16.09 (24.90)	16.97 (5.47)	23.87 (40.66)	22.97 (-3.77)	22.97 (0)
4.	No. of milk Product Factories	2	5 (150)	6 (120)	6 (0)	7 (16.67)	8 (14.28)	12 (150)
5.	No. of Feed Mixing Plants	4	6 (150)	6 (0)	7 (16.67)	7 (0)	7 (0)	7 (0)

Source: Department of Planning and Development, APDDCF Hyderabad, 1984.Compiled from the records of NDDB Annual dairy progress reports on Operation Flood different Plan documents, Finance and Planning Department, Government of Andhra Pradesh. Records of APDDCF, Hyderabad, 2002.

day in 1996-97 which continued to be same and same is seen in 2000-01. The number of milk product factories increased from 2 to 12 between 1975-76 and 2000-01 and the number of feed mixing plants increased from 4 to 7 during the same period. The analysis of the table reveals that there has been satisfactory growth in the infrastructural facilities. But there has been low utilization of capacity during the period, leading to turnover and low profitability.

Bovine Population in A.P.

The State, however, has earned all-India fame for its 'Ongole' cow which is a high milk-yielding variety. The State is known to possess dual purpose breeds other than Ongole, like Malhi and Deoni. These breeds of cows are also known to give milk of high fat content. The Punganur and Kalikiri breeds are extensively found in Chittoor district. In Kurnool district, Doopadu; a short-structured breed, is reared. In Coastal Andhra, Mahaboobnagar, Nalgonda and Khammam districts "Krishnaloya' type animals are breed for milk. In Telangana and Medak districts are known for Deoni breed. Though in Telangana malvi breeds are generally found, the 'Murrah' graded buffaloes are mainly breed for milk. They are mostly found in Hyderabad, Vijayawada, Mandapeta and Rajahmundry.

Table 1.14 reveals that the growth of cattle and buffalo population in the State between 1951-1999 was meager 1.57 per cent per annum. The number of cattle decreased by 10.67 per cent witnessed negative growth rate at 0.79 per cent per annum compensated by the growth of buffalo population, at 5.14 per cent.

Table 1.15 shows that the trend in the bovine population in relation to human population and operational holdings. It is disheartening to note that the total bovine population in relation to human population and operational holdings has been decreasing consistently. The decline is more pronounced in relation to operational holdings. Within cattle population, the male cattle in relation to human population

Table 1.14. Cattle and Buffalo Population in A.P.

Particulars	1951	1956	1961	1966	1972	1977	1982	1987	1993	1999	Compound growth rate	't' value
Cattle												
Males (Over 3 years)	5007	4842	5279	5403	5491	5401	5190	5420	5249	4902	0.21	0.39 NS
Females (Over 3 years)	3924	3762	4075	4154	4231	4114	4390	3564	3030	2930	.2.82	2.29 NS
Young stock	3318	2672	2991	2785	2785	2526	3640	3390	2668	2738	0.25	0.17 NS
Total	12249	11276	12345	12342	12507	12041	13220	12374	10947	10570	0.79	1.07 NS
Buffaloes												
Males (Over 3 years)	1362	1102	1402	1460	1275	1275	1120	922	665	632	.7.85	3.84 **
Females (Over 3 years)	2942	2876	3151	3219	3668	3668	4327	4475	4807	5200	8.37	5.69 **
Young stock	2279	1990	2396	2112	2215	2219	3257	3360	3670	3807	7.47	4.99 **
Total	6583	5968	6949	6791	7057	7162	8704	8757	9132	9639	5.14	7.89 **
Grand Total	18832	17244	19294	19133	19564	19203	21924	21131	20079	20209	1.57	2.89 *

Note: NS: Not significant at 5 % level.
* Significant at 5 per cent level.
** Significant at 1 per cent level.

Source: Compiled from different livestock census reports, Government of A.P., Directorate of Economics and Statistics, Hyderabad. *Statistical Abstract of A.P.* 2000, p. 124.

Table 1.15. Category-wise Trends in Bovine Population: From 1985 to 2002

Category	Bovine population per 1000 human population					Bovine population per 1000 operational holdings				
	1985	1987	1993	1997	2002	1985	1987	1993	1997	2002
Male cattle (Over 3 years)	97 (37.74)	101 (43.72)	79 (47.87)	76 (42.45)	71 (40.57)	708 (39.46)	658 (43.78)	596 (46.52)	615 (22.01)	605 (38.00)
Female cattle (Over 3 years)	92 (35.79)	67 (29.00)	46 (27.87)	65 (36.31)	68 (38.85)	594 (33.11)	433 (28.80)	403 (31.46)	548 (35.17)	562 (35.30)
Young stock	68 (26.40)	63 (27.27)	40 (24.24)	38 (21.22)	36 (20.57)	492 (27.42)	412 (27.42)	282 (27.41)	395 (22.01)	425 (26.69)
Total cattle	257 (100.00)	231 (100.00)	165 (100.00)	179 (100.00)	175 (100.00)	1794 (100.00)	1503 (100.00)	1281 (100.00)	1558 (100.00)	1592 (100.00)
Male Buffaloes (Over 3 years)	21 (12.88)	17 (10.36)	9 (6.56)	7 (5.42)	4 (3.27)	149 (12.61)	112 (10.52)	42 (4.95)	79 (7.72)	72 (6.73)
Female Buffaloes (Over 3 years)	81 (49.69)	84 (51.22)	73 (53.28)	70 (54.26)	68 (55.74)	590 (49.96)	544 (51.12)	464 (54.71)	508 (49.65)	535 (50.04)
Young stock	61 (37.42)	63 (38.41)	55 (40.14)	52 (4031)	50 (40.98)	442 (37.42)	408 (38.34)	342 (40.33)	436 (42.62)	462 (43.21)
Total Buffaloes	163 (100.00)	164 (100.00)	137 (100.00)	129 (100.00)	122 (100.00)	1181 (100.00)	1064 (100.00)	848 (100.00)	1023 (100.00)	1069 (100.00)

Source: Government of A.P., Summary Report on Livestock Census, Directorate of Economies and Statistics, Hyderabad, 1997, 2002.

and operational holdings have been declining more rapidly than female cattle and the young stock. This is because of declining use of male cattle as drought power animals. The decline is very significant in the case of male buffaloes; there are marked difference in the shares of crossbreed and indigenous varieties, in total cow population, which can be seen from Table 1.16. The distribution of bovine population shows large variations among districts. The indigenous cows outnumber than cross-breed variety. Which is concentrated in Chittoor district (61.78%) followed by Srikakulam (6.93%) and East Godavari (4.33%). Indigenous varieties of cows are widely distributed in all the districts. Nevertheless, the districts of Mahaboobanagar (9.20%), Adilabad (8.00 percent), Khammam (7.87%) and Nalgonda (6.74%) have greater percentage of indigenous cows. As far as buffaloes are concerned Guntur district has the highest percentage (9.03%) It can be noticed that buffaloes are concentrated in Coastal region, whereas indigenous cows in Telangana and crossbred in Rayalaseema.

With regard to production of milk, Guntur district ranks high (8.36%) followed by West Godavari (7.94%), Chittoor (7.88%) and Krishna (7.11%). Confined reading of distribution of bovine population and the milk production reveals that West Godavari occupying second place in the production of milk, has relatively low percentage of bovine population. But the districts of Guntur and Chittoor have relatively greater percentage of bovine population than justified by their relative shares in milk production.

Milk Production and Per Capita Availability of Milk

Dairy development can be understood in terms of trends in production of milk and the per capita availability of milk, during development process along with the bovine population, productivity and the total milk production, there should be an increase in per capita availability of milk at least in average terms. A.P. possesses large number of milch

Table 1.16. District-wise Population of Cows and Buffaloes and Milk Production 1996-97 to 2001-02

Districts	Crossbreed cows (in'000) 1996-97	Indigenous cows (in'000) 1996-97	Buffaloes (in'000) 1996-97	Milk production (in'000 tonnes) 1996-97	Milk production (in '000 tonnes) 2001-02
1	2	3	4	5	6
Adilabad	0.5 (0.25)	173.9 (8.00)	83.8 (1.88)	NA	NA
Anantapur	6.1 (3.00)	144.9 (6.66)	154.3 (3.47)	119 (3.16)	95 (2.5)
Chittoor	125.6 (61.78)	141.9 (6.53)	89.6 (2.01)	297 (7.88)	305 (8.3)
Cuddapah	2.9 (1.44)	33.8 (1.56)	229.9 (5.17)	119 (3.16)	89 (2.4)
East Godavari	8.8 (4.33)	83.6 (3.84)	254.7 (5.17)	247 (6.56)	252 (6.8)
Guntur	1.2 (0.59)	22.9 (1.05)	401.5 (9.03)	315 (8.36)	310 (8.4)
Hyderabad	0.6 (0.29)	0.8 (0.40)	9.20 (0.21)	16 (0.420)	15 (0.4)
Khammam	0.3 (0.14)	171.0 (7.87)	225.4 (5.07)	NA	NA
Karimnagar	3.4 (1.67)	90.9 (4.18)	176.0 (3.96)	189 (5.02)	180 (4.9)
Krishna	3.4 (1.67)	34.3 (1.58)	360.6 (8.11)	268 (7.11)	276 (7.5)
Kurnool	1.0 (0.49)	107.9 (4.96)	226.5 (5.09)	144 (3.82)	124 (3.4)
Mahaboob Nagar	4.6 (2.26)	199.9 (9.20)	159.7 (3.59)	139 (3.69)	129 (3.5)
Medak	1.4 (0.69)	87.5 (4.03)	133.0 (2.99)	127 (3.31)	122 (3.3)
Nalgonda	2.8 (1.38)	146.6 (6.74)	258.6 (5.87)	161 (4.27)	158 (4.3)
Nellore	0.8 (0.39)	70.8 (3.26)	294.2 (6.62)	196 (5.20)	206 (5.6)

1	2	3	4	5	6
Nizamabad	1.2 (0.59)	62.6 (2.88)	122.9 (2.76)	94 (2.49)	90 (2.4)
Prakasam	0.8 (0.39)	36.2 (1.67)	384.6 (6.62)	240 (6.37)	245 (6.6)
Ranga Reddy	4.6 (2.27)	65.2 (3.00)	109.0 (2.45)	140 (3.72)	130 (3.5)
Srikakulam	14.1 (6.94)	16.00 (8.97)	55.8 (1.25)	76 (2.020)	72 (2.0)
Visakhapatnam	6.1 (3.00)	69.7 (3.21)	186.1 (4.19)	138 (3.66)	132 (3.6)
Vijayanagaram	6.3 (3.10)	102.5 (4.71)	102.6 (2.31)	95 (2.31)	90 (2.4)
Warangal	1.4 (0.68)	105.7 (4.86)	173.9 (3.91)	110 (2.92)	101 (2.7)
West Godavari	5.4 (2.65)	60.0 (2.76)	255.6 (5.76)	299 (7.94)	310 (8.4)
Total	203.3 (100.00)	2173.5 (100.00)	4445.0 (100.00)	3766 (100.00)	3685 (100.00)

Note: Figures in parentheses are percentages to respective column total.
NA: Not available

Source: Dr. V.K. Pandey, Livestock Economy of India, *Indian Journal of Agricultural Economics,* Vol (3), 1997. *Statistical Abstract,* 2002 A.P.

animal population, but the State lags behind in milk production. This is mainly due to low productivity of the non-descript animals. In certain areas cattle are reared mostly for draught purpose and manure. In some districts cattle population is more in numerical figures but they are not habituated for milk. Table 1.17 shows the milk production and the per capita availability of milk in A.P. It is a revelation that wide gap between growth in milk production and per capita availability and to the fact that production has been lagging with the behind the growth in the population. Between 1997-98 and 2006-07, the total production of milk increased from 4.47 million tonnes to 7.94 million tonnes, the registering annual average growth rate of 4.33 per cent. The per capita availability of milk increased from 167 grms per day to 269 grams per day recording 2.19 per cent of compound growth. The preceding analysis reveals that, the growth of per capita availability has been far less than the growth of milk production.

Table 1.17. Milk Production and Per Capita Availability in Andhra Pradesh

Year	Milk production (in million tonnes)	Growth rate over previous year % age	Per capita availability (in g/day)	Growth rate over previous year % age
1997-98	4.47	...	167	...
1998-99	4.84	1.08	185	1.10
1999-2000	5.12	1.05	192	1.03
2000-01	5.52	1.07	194	1.01
2001-02	5.81	1.13	209	1.07
2002-03	6.58	1.05	231	1.10
2003-04	6.96	1.06	238	1.05
2004-05	7.26	1.04	250	1.04
2005-06	7.62	1.05	260	1.03
2006-07	7.94	1.04	269	1.03
Avg.	6.21	-	219.5	-
CGR	6.44	-	5.21	-

PROCUREMENT AND SALE OF MILK

The procurement and sale of milk are the two important aspects of marketing of milk which makes the activity of dairying more profitable to the milk producers. Proper marketing is necessary for milk and milk products because of their perishable nature. It also acts as a catalyst for rapid development of the dairy. Procurement and sale of milk depends upon the infrastructure and the market net-work more so in the rural areas.

Table 1.18. Procurement and Sale of Milk in Andhra Pradesh

(in lakh litres)

Year	Procurement	Milk sales	Sales as percentage of procurement
1997-98	3112.42	2273.00	73.03
1998-993	282.66	2386.44	72.70
1999-2000	3431.22	3329.56	97.04
2000-01	3723.91	3264.38	87.72
2001-02	4200.48	3311.76	78.85
2002-03	4535.25	3575.35	78.82
2003-04	4685.50	3765.48	80.36
2004-05	4960.44	3960.80	80.00
2005-06	5250.65	4200.50	80.00
2006-07	5445.45	4568.54	83.90
Avg.	4262.80	3463.58	
CGR	6.66	6.69	-

From which most of the milk production comes. Most of the milk is purchased and sold in the unorganized market.

Table 1.18 shows procurement and sale of milk from 1997-98 to 2006-07. It can be seen from the table that the milk procurement from 3112.42 lakh liters in 97-98 to 5445.45 lakh litre in 2006-07. There seems to be wide fluctuations in milk procurement throught the period. A similar trend is obtained in milk sales the excess of milk sale over procurement was due to reading the fat from the standard level of 4.5 per cent to either 3.0 per cent or 1.5 per cent. In the years when procurement of milk was greater than sales, excess milk was converted into milk products of greater durability.

Conclusion

Share of livestock sector hovers round 5 per cent of GDP of India. Bovine population constitutes 58 per cent of livestock population in India in 2002. India with 19 per cent share in world milk production closely followed by U.S.A. with 16 per cent share. Yearly percentage changes in milk production have been constituently positive. Per capita availability of milk in India was 231 grams per day in 2004. Anand model of cooperative milk cooperative structure integrates rural producers and urban consumers. Replication of model led to phenomenal growth of societies and membership. OF-I, II, III achieved unprecedented growth in procurement, processing, and marketing of milk and provision of technical inputs, percentages share of outlay on animal husbandry and dairying in total plan outlay was 0.11 per cent .in first plan and 0.33 per cent in eighth plan. Due to external sector liberalization dairy industry has turned to be highly competitive with the entry of private players, domestic and foreign. Perceptive 2010 aims at beginning more women into cooperative field.

Share of primary sector was 32 per cent in SGDP and 69 per cent in employment in 1999 to 2000. APDDC was established in A P with the objective of bringing socio-economic transformation with massive dairy development programs. Female cattle and buffaloes per 1000 population and 1000 operation land holdings have kept increasing in 1980s and 1990s. District wise analysis reveals that Chittoor topped the list with 80.1 per cent share in total milk production in 2001-02 per capita milk availability was 147 grams per day in 2001-02, which was above the national figure. Procurement and sales of milk registered CGRs of 4.22 and 3.79 per cent respectively during 1980-02 in A P.

REFERENCES

1. *Economic Review* 2004, State Planning Board, Thiruvantapuram, 2005, p. 78.
2. Mehta, Pradeep S. "Indian Dairy Sector: The Challenges Ahead," *Indian Dairyman*, Vol. 56 No. 10, 2004, p. 153.
3. Ravisankar "Policy and Regulation – Influencing Milk Producers", *Indian Dairyman*, Vol. 55, No. 3, 2003, p.100.
4. George, P.S. and K. N. Nair, Livestock Economy of Kerala, Trivandrum, Centre for Development Studies, 1990, p. 5.
5. Reddy, K. P., "Initiatives for Achieving Excellence: Indian Dairy Industry Scenario," Indian Dairyman , Vol. 56, No. 10, 2004, p. 46.
6. Kulandaisamy, V., 1986.Co-operating Dairying in India, Rainbow Publications, Coimbattore. p. 6.
7. Jain, J.L. "Organised Milk Marketing IN India Socio-Economic Impact", Kumarappa Institute of Gram Swaraj, B-190, University Marg, Bapunagar, Jaipur-302015.
8. Agarwal, Dr. V.K., Marketing of Dairy Products in Western UP, Himalaya Publishing House, Bombay.
9. Bhanj, S.K. and Hema Tripati.2004. "Strategic Interventions through Dairying for Rural Development", Journal of Rural Development, Vol. 23(1), p. 83.
10. Gnana, C., 2004. Production and Marketing of Aavin Milk and Milk Products: A Study, Tamilnadu Journal of Co-operation, p. 10.

11. Ravisankar (1997), Case in India, The Case of Cooperative Dairying in India- Report on an International Workshop at the International Institute, Histadrut, Betberi, Israel, COOPNET, p. 42

12. Pichai, C. & R.Selva Raj (1999), India's Production of Milk and Milk Products, Vis-à-Vis the World Dairy Scenario, Tamilnadu Journal of Cooperation, Vo!. 91, No. 5, P.26.

13. Wright, N. C., Director of Dairy Research Institute, Scotland, 1984. Milk Procurement and Technical Inputs Manual, Anand, NDDB, p. 13

14. Mascarnnas, R.C.A Strategy for Rural Development-: "Diary Cooperatives in India", Sage Publications, New Delhi 1988, p. 77.

15. Low, Allex Laid, The Maritime Cooperative, July 1975, Reproduced in The Anand Pattern of Cooperative Development: Kaira Milk Producers Cooperative Union, Anand Mascaren R.C., O.P. CIT, p. 78.

16. National Dairy Development Board, Anand Pattern and Operation Flood: An Overview, Anand , NDDB 1985, p. 27.

17. Khurody, D.N."Dairying in India" Asian Publishing House, Bombay.1974, p. 41.

18. Benargy, A., Dairying Systems in India, http: www// fao.org, p. 1.

19. Vidyanathan, A., Bovine Economic in India, Trivandrum, Centre for Development Studies, 1988, p.,1.

20. Bedi, M.S., Diary Development and Marketing and Economic Growth, New Delhi, Deep and Deep Publications, 1987, p. 9.

21. Gopalakrishnan, C.A. and G. Morli Mohan lal "Livestock and Poultry Enterprises for Rural Development". New Delhi, Vikas Publishing House, Pvt. Ltd., 1984, p. 4.

22. Birthal, Pratap, S., "Technological Change in India's livestock Sectors and Its Impact ", Livestock in Different System in India, Agricultural Economics Research Review (Conference Proceedings), Advance Publishing Concept, New Delhi, 2002, p. 18.

23. Khanna, R.S., Sustainable Development of the Dairy Industry in Asia: Challenges and Opportunities "Indian Dairyman, Vol. 57, No. 2, 2005, p. 26.

24. Reddy, K.P., "Initiatives for Achieving Excellence Indian Dairy Industry Scenario", Indian Dairyman, Vol. 56, No. 10, 2004, p. 46.

25. Bedi, R.D., 1958" Theory, History and Practice of Cooperation" Meerut Loyal Department , p. 23.

26. Cole, H H and Manager Rouning, 1980, Animal Agriculture, Freeman and Company, Sanfrancsco, p. 70.

27. Sanath, K. A. 1973" Milk Chilling Centers and Management Indian Dairy Holder Association, A.P, Hyderabad, pp. 1-3.

28. Kulandhaisamy, V. Cooperative Dairying in India, Rainbow Publication Coimbatore 1986, p. 6.

29. Varma, Dr. Madhu Sudhana, "Production and Marketing of Milk and Milk Products" SV Press, Tirupati, 2004.

30. Sharma, Dr. PSRVKP, "Dairy Scenario in Marketing" Sharma Society, Rajahmundry, 1995.

31. Five Year Plan 1978-83, Andhra Pradesh, Draft on livestock, Agricultural Department, Government of Andhra Pradesh, Hyderabad, p. 88.

32. Inaugural Souvenir of Milk Products Factory at Hyderabad, "Andhara Pradesh Dairy Development Corporation", Hyderabad, 1975, p. 5.

33. Nageswara Rao, T. "Twenty Five years of Andhra Pradesh", Telugu Academy, Hyderabad, 1980, pp. 73-74.

34. Five Year plan, 1978-83, Finance and Planning Department, Government of Andhra Pradesh, 1978, p. 88.

35. Second Five Year Plan 1956-61, Finance and Planning Department, Government of Andhra Pradesh, pp. 201-208.

36. Third Five Year Plan 1961- 66, *op. cit.,*, 1961, pp. 175-183.

37. Fourth Five Year Plan 1969-70 to 1973-74; *op. cit.,* 1969, p. 57.
38. Fifth Five Year Plan 1978-83, Andhra Pradesh, Draft out line, op.cit., November, 1978, p. 88.
39. Sixth Five Year Plan 1980-85, *op. cit.,* 1980.
40. Seventh Five Year plan 1985-90, *op. cit.,* 1985.
41. Eighth Five Year Plan 1992-96; *op. cit.,* p. 57.

2 CHAPTER

REVIEW OF LITERATURE AND RESEARCH DESIGN

INTRODUCTION

In this chapter an attempt is made to review of literature pertaining to dairy industry with specific focus on marketing practices. Thus, Part-A is exclusively devoted to review of literature. Part-B of this chapter presents statement of the problem, need for and significance of the study, objectives, sample, data sources, tools of analysis, scope and limitations of the study and organization of the thesis.

PART-A: REVIEW OF LITERATURE

A thorough review of literature of theoretical as well as empirical works in dairy industries in general and marketing practices in particular helps us in settings objectives and broader directions of the study. Innumerable studies were conducted in the area of dairy industry, lent a few select studies are reviewed hereunder under broad categories captions. Sufficient care is taken to included both foreign and Indian research studies so as to make the review comprehensive, analytical and critical, keeping in view the underlying theme of present research. Hence, there is a common thread wearing all the studies and providing broader framework and avoiding over/under laps.

Studies on Marketing of Dairy Products

The following studies a selected to making of dairy products.

Girdhari[1] underlined the need for developing an effective marketing organizations for dairying which results in creating suitable systems viz procurement, process, pricing, packaging and distribution of milk and milk products.

David Avery Vosse[2] observed that the structure of procurement, market for fluid milk served by a single co-operative society approximated to monopoly, while markets having more sellers resembled a highly concentrated oligopoly wherein the services constituted a source of product differentiation. Sanitary regulations, milk marketing orders, full supply contract, perishable nature of product and transportation costs influenced the strength of barriers to entry.

Shaik[3] studied milk marketing practices and opined that marketing was not given important in dairy industry. Professionalism was absent in the marketing of dairy products. The author advocated need for attitudinal change especially when one considers the future potential of growth of dairy industry.

Mahinder Kaur and Gill[4] examined the system of marketing of milk in terms of channel cost and profit margin in Ludhaiana district of Punjab. The authors concluded that the direct channel i.e. (producer-consumer) was the most efficient from the point of view of producers as well as consumers.

Sukumar[5] in his book 'Outlines of Dairy Technology' stated that milk is necessarily produced with in a short distance from the consumption centre because it is perishable in character. Incentives like protection of milk during transportation, reasonable price for milk and milk products, and modern knowledge of marketing should be offered to augment the supply-milk to consumers.

Kalra and Rajvirsingh[6] in their study found that the maintaining system had created linkage between milk producers in rural area and consumers in the distant urban areas. Further, they pleaded for an efficient marketing system to transform the socio-economic life of rural people.

Patel[7] in his study concluded that ready and remunerative market for milk and milk products had improved the economic conditions of many milk producers.

Jawana Ram[8] made a case study on the marketing of milk and milk products in Rajasthan. Tetra-pack milk, butter, ghee and skimmed milk powder were marketed through three kinds of channels of distribution. The first channel was concerned with the direct sale of milk to consumers. The second channel i.e. retail agents, were engaged in both milk and milk products, and in the third channel wholesalers were engaged for distributing milk products.

Chahal and Balwinder Singh[9] reported that the milk co-operative societies were the most important marketing channel to their members in Punjab. The percentage of functional milk societies increased from 17.61 in 1980 to 93.15 in 1990. This phenomenal growth of milk co-operative societies resulted in efficient marketing facilities and reduced the exploitation by private competitors.

Mandanna and M.V. Srinivas Gouda[10] in their study reviewed the features and problems encountered by dairy industry in India. Milk marketing was represented by unorganized private traders who turned milk trading into an exploitative market. Now-a-days, for marketing of milk it is felt that well conceived and organized net-work of dairy co-operatives right from village level would be essential for the speedy growth of the dairy industry in the country.

Sah[11] in his study concluded that the Indian dairying was currently in a state of transition with the rapid creation of milk marketing facilities benefiting rural milk producers and also the under-privileged urban consumers. A wide range of dairy products are marketed such as better, cheese, baby food, and skim/whole milk powder. Once imported were now produced in the Country, which was due to strengthening of the milk marketing facilities. The marketing of milk powder had increased from 22,000 tonnes in 1970 to 1,95000 tonnes in 1994.

Prabaharan[12] demonstrated the impact of marketing of live-stock products at remunerative prices on the living conditions of rural people. The recent economic reforms and liberalization are expected to open up accelerated market-led opportunities for live-stock products.

Surya Murthi[13] in his study suggested strategies for the marketing of milk which include consumer satisfaction; a uniform strategy for quality control through a well structural and planned institutional network; strengthening co-operatives involved at the village level in milk-procurement network; controlling product cost by minimizing the overheads; improving market competitiveness of dairy units by handing over them to specialists in each area of production, procurement, processing and marketing. Fair and equitable competition is vital for dairy industry. Co-operatives have to develope expertise in managing the politico-legal environment.

Kumar[14] evaluated the procurement and distribution of milk by District Cooperative Milk Producers' Union Ltd Villupuram. Credit sales increased as against in cash sales.

Cost surpassed the price and resulted in loss throughout the study period. The results of a study conducted by *Pitchai*[15] showed that the quantum of milk procurement depended on monsoons and the price offered by private traders to co-operatives. He compared the cost increases in four unions. The production cost and sales price were Rs.7.31and7.79 in Salem union, when compared to those in Erode (Rs.7.44 and Rs.7.91) all the unions-supply milk to the federation at Rs.6.48 per liter, which was less than the cost. Transportation cost was the major component in the distribution cost of milk. The "t" test revealed that there was a significant difference in the average monthly expenditure on milk and milk products across the income groups in the same town but the difference was in significant between the same income groups across the towns. A majority of non-buyers of Aavin products reported that required quantity was not available. The procedure of lodging complaint in Aavin was uneasy.

To find out the performance of co-operatives in the marketing of the milk, *Chahal*[16] Organized a study in Punjab. The milk was sold to co-operative societies, milk collection centres, milk contractors, milk vendors, sweet shops and local customers. The study pointed out that the percentage of milk sold to milk vendors had decreased with the increase in farm size. With increasing marketable surplus milk producers curtailed the sales to milk vendors and preferred sweet shops. This was because of price differentiate. Further, the author viewed there was a need to strengthen the milk co-operatives so as to make milk availability a successful venture. This would go a long way to help the dairy farmers to get remunerative prices for their produce, and consumers in turn be benefited by way of providing quality milk at reasonable price.

Sanghu[17] studied the production consumption and marketed surplus of milk in Meerat district of Uttar Pradesh. The study focused on 193 producers with 444 milk animals spread over different categories of milk producer households. The contribution of medium farmers with buffaloes in the milk production was the highest, when compared to landless and small farmers. It observed that there was a negatively association between land holding size and marketed surplus.

Jairath et al.[18] in their study on Mandi Milk Supply Scheme of Himachal Pradesh reported that there were lean and flush seasons in the procurement of milk. The index of arrivals reached its peak season during January and thereafter declined.

Manohar and Sudarshan[19] analyzed the role of dairy development and other agencies in the development of dairy units, employment opportunities and the livelihood of the weaker sections of the selected sample societies in their study area.

Patel and Prabaharan[20] conducted a study with 300 households in Chennai. The study concluded that it would be beneficial to the Tamil Nadu Dairy Development Corporation if active steps would be taken to infuse knowledge about various important dairy processes and services to the consumers. A suggestion was also made to explore the possibilities of supplying pure cow milk to the consumers in the city as the preference for cow milk was more compared to buffalo milk. It was also suggested that a small pocket of milk (250ml) might cater to needs of poor and low-market segments.

Chinnaiyan et al.[21] organized a study on market functionaries, market channel and price structure in Erode taluk of Tamil Nadu. Market structures were identified as oligopolistic at the assembling as well as in distribution levels. Further it was found that the unorganized sector held a large share of milk market and even sent milk outside Tamil Nadu to places like Bangalore and Mangalore.

Ravikiran et al.[22] studied the milk production in Krishna District of Andhra Pradesh. The study concluded that buffaloes formed the main-stay in milk production in Krishna District. Krishna Kamadhenu scheme under which crossbreed cows were brought from Karanataka and supplied to farmers for better yield, resulted in the collection of more milk to the society and union. *Gupta and Devaraj*[23] conducted a study on Churu District of Rajasthan. The study revealed that the proportion of milk retained for household consumption was as high as 73 per cent marketed surplus was estimated at 27 per cent this is on account of inadequate marketing facility. The marketed surplus was the lowest in the case of landless labourers. Apart from its utilization in fluid form, a significant proportion of milk was used for ghee.

STUDIES ON ECONOMICS OF DAIRYING

Patel and Pandey[24] in their study on the "Economic Impact of Kaira District Co-operative Milk Producers' Union" highlighted the role and functioning of dairy co-operative, milk yield per animal, household dairy income and per capita milk consumption. By way of comparison the data analysis and results were presented 100 respondents each ordered from controlled and non-controlled groups of villages by comparing 100 respondents of controlled and non-controlled villages selected from the study area. Studies on economics of dairying lay emphasis on cost and returns and the profitability of dairy industry in many parts of the country.

A number of economists and researcher like Gunnar *Myrdal,*[25] *Mishra,*[26] *Ramachandran*[27] attempted to show that India has surplus cattle. They insisted on a strategy for enhancing milk production by improving the quality of milch animals without adding to the number. But these studies failed to provide any estimate of such surplus in milch animals. An alternative approach adopted by researchers has been the study of economic relationships at the firm level or the economic impact of dairying.

Grill and Grill[28] analyzed the role of dairying in an increasing the income levels of people below poverty line. They noted that regular and evenly distributed income from dairying throughout the year brought a significant change in the traditional economic condition of farmers who were always in debts in order to meet unavoidable routine expenses. Their study also reported that dairying helped in raising the living standards of milk producers.

Suhag et.al.[29] showed that livestock enterprise had a positive effect on the income and employment generation in all farm sizes.

Melville[30] emphasized that to develop dairying as an instrument of change; small farmer must sell his surplus milk every day at remunerative prices.

Acharya, and Pawar[31] reported that the average daily and total milk production per lactation of cross breed cows was 8.68 litres and 2,609 liters respectively. The respective figures for buffaloes and local cows were 4.13 and 1,359 liters and 2.26 and 604 litres respectively. It was found that the per day per animal labour utilization was the highest in cross-breed cows, at 2.02 hours as compared to 2.53 hours in a buffalo, and local cow. It was shown that profitability more in crores breeding programme than in traditions local cows.

Rangarajan et.al.,[32] analyzed the economics of milk production from cross-breed cows and nondescripts in terms of yield, cost of milk production, net income realized by the farmer, employment generation and efficiencies resource were in rural areas of Periyar district of Tamil Nadu. The main finding of the study was that cross-breeds fared far better than local ones in all the aspects.

Sharfuddin[33] examined the impact of dairying on income and employment and examined whether organized dairy had any economic importance when compared to unorganized dairy. The researcher concluded that the organized dairy direct and significant impact on employment generation, milk production and return for weaker sections of the society.

Bhanja and venkatadri[34] to assess the role of Milch Animals under Integrated Rural Development Programme (IRDP) a survey was conducted by Bhanja and Venkatadri in agriculturally developed and backward States of Haryana and Orissa. The study concluded that the net income and employment generation through the programme was than in Haryana.

Biraday[35] conducted from his study that the dairy development had a decisive impact on beneficiary groups' i.e. landless labourers, marginal, small and medium farmers than on large farmers.

Kirshnama Raju and Seshaiah[36] reviewed dairy development in the country during planning period and also analyzed the causes for its slow progress In their opinion the primary reason for the slow progress of dairying was the lack of proper coordination among the different institutions involved in dairy development.

Mitra[37] organized a study on the impact of Operation Flood (OF) programmes on rural women milk producers. More specifically he attempted to provide insights into the extent of women involvement in dairy production and management. He observed that for women of agricultural households, the intensification of dairying meant long and more working time also a day. It meant involvement of their children, particularly daughters, in various kinds of work which led to withdrawal of girls from school. He felt that the establishment of cooperatives and the consequent generation of cash income might deprive women of the control over the income from dairying.

Geeta Somjee[38] studied the nature and extent of change in various aspects of socio-economic life of rural communities as a result of their exposure to the organization, principles and activities of milk cooperatives at Amul in Kaira district, Dudhsagar in Mehasana district, Surnorl in Surat district, and Sabar in Sabar Kantha district of Gujarat. The findings of study revealed that their insistence on joining a common queue in front of the dairy cooperatives helped to eliminate caste hierarchy in the Dairy Cooperative Societies (DCS), helped in increasing women's participation.

National Council of Applied Economic Research (NCAER)[39] observed that 'substantial gains' had accrued to the milk producers, particularly the economically weaker sections of the rural population due to OF covering nearly 61,600 village cooperative societies and procuring on an average 90 lakh liters of milk a day to meet the growing urban demand for liquid milk. The study found that 43 per cent of sample households had only one milch animal and almost 75 per cent of the members belonged to the landless laborers, marginal and small farmers.

Sekhar Dutt[40] pointed out the shortcomings of OF-II and offered suggestions to overcome them. He favored dovetailing OF-II and antipoverty programmes for effective utilization of animal wealth and man power.

Thirunavvukkarasu et al.,[41] observed that the benefits of 'OF' were more skewed towards upper castes than the lower castes. They further observed that special efforts were needed to remove the constrains faced by scheduled tribe households in fully exploiting the potentials of the OF.

Studies on Women Participation in Dairying

Somjee and Somjee[42] reported that managerial shift had taken place in dairy industry and Women had changed their outlook, broken the barriers of seclusion from men, and took managerial role in dairying.

Mitra[43] in his study found that landless women producers kept fewer animals than others. Further they spent two and a half hours daily on hard physical work like cleaning, washing, fetching grass, feeding cows, milking and sale of milk. But women from small farmer house hold with large livestock holdings spent an hour or more on dairy activities.

Nageswara Rao[44] in his study found that in rural areas women were primarily engaged in various chores connected with dairy. To ensure clean and quality milk production, women participation was necessary. Women who occupied half of the worlds population perform two-thirds of the worlds dairy work, received one-tenth of its income and own less than one hundredth of its property. Women co-operatives helped in bringing women into main stream of life. Now-a-days, State government started the taking various steps to start co-operatives in rural areas to encourage the participation of women in dairies.

Mascaren[45] in a study concluded at that animal husbandry and dairy should have women laborers. Most of the housewives in rural areas earned additional income from dairying.

Sharma and vanjani[46] with reference to Shankpur village in Rajasthan, observed that women went twice a day to fields to collect and carry huge headhoads of fodder to their milk animals. They not only fed their milk animals twice a day but also cleaned the shed. Further, poor women with buffaloes spent 3.5 to 4.5 hours daily in milk production and related activities daily.

Kulandaisamy[47] in his study concluded that the livestock and dairy had been one of sectors where participation of women work force was high. Poor rural women performed a large part of work relating to maintenance of dairy cattle, milk production and processing. The women's welfare department had agreed to sanction 20 per cent of money for purchase of milk cattle by women co-operatives members.

Subramanyam[48] had emphasized that with organized milk procurement through co-operatives was an attempt to organize poor rural households and poor women as members of co-operatives and to train them in scientific methods of breeding, cattle care and milking. Poor rural women had purchased milk animals with the institutional assistance and, thereby increased income and the standard of living.

Mudgal[49] in his study found that men looked after all agricultural activities, whereas women engaged in taking care of animals. It were women at home who took care of animals and develop an unseen bridge of love and affection with animals. They understood animals better. He must exhorted that e must tap this situation for improving cattle health, hygienic milk production and overall development of the society.

Studies on Bovine Population and Milk Yields

A few studies were carried out on the bovine population and yield of milk.

Gangwar[50] found that there were inter-district variations in bovine population, cost structure, production and disposal of milk. The study revealed that in the cattle population 50 per cent was bovine population. Buffaloes were more in wet districts on compared to bovine population.

From his study *Kumar*[51] arrived at two major conclusions. Firstly there was a positive and significant relationship between feed and milk yield. Finally production elasticity's of feed were higher for the third and fourth-order of lactations, implying that dairy farmers had a greater scope of increasing milk production through manipulation of levels of feed ratios for their cows.

Shah and Sharma[52] studied the supply functions of milk and showed that the milk yield of cross-breed cows was higher and its potentiality increased through crossbreeding, feeding and management. The milk supply could be increased substantially by proper feeding of cross-breed cows.

Studies on the Consumption and Availability of Milk

A number of studies were organized on the aspects of availability and consumption pattern of milk. *Gupta*[53] estimated that the per capita consumption of milk was 140 grams in India while in European countries it was between 1990 to 1698 grams per day. Further 43 per cent of milk available was consumed in liquid form and the remaining was converted into kova, dahi etc.,

Sharma et al.[54] worked out that the economic demand for milk at 36.68 million litres as per recommended level of 210 grams per capita per day. According to this study, the milk production should increase at the rate of 3 metric tonnes, annually up to 1978, and thereafter at the rate of 4mt, to bridge the increasing gap between the demand and availability.

Gupta and Pandey[55] projected the demand for and supply of milk in India for the years 1979, 1984 and 1986. They estimated that the gap between demand for and supply of milk in the country was at 4.5 mts on the assumption of 250 grms per capita consumption per day.

Shanty George[56] that enhanced milk production through observed the dairy cooperatives from rural areas. The milk producers have readily sold their milk to dairy cooperatives. Which adversely affected the milk consumption by the rural masses in general and milk producers in particular.

Studies on Capacity Utilization and Production

Kumar and Ranf[57] made an attempt to study the effect of breed, lactation period, maintenance cost etc., on milk production in Haryana. The authors found that the feed components significantly influence the milk yield.

Sankayam and Joshi[58] organized the study to establish input/output relationship in indigenous and crossbreed cows. Two types of production functions, namely linear and Cobb-Douglas were used to express the relationship between milk output per milch animal and various factors influencing it. They found that the feed-stock significantly influenced milk output in cross-breed cows compared to indigenous cows.

Muranjan[59] identified the key factors affecting the milk procurement viz., procurement price, price of related commodities, growth of procurement agencies and changes in the overall productions of milk in public sector dairy plants in Maharastra.

Madhavan[60] held that the success of a dairy plant would depend on the organization of milk procurement system. Further, dairy plants suffered due to under-utilization of installed capacity. Under utilization was mainly on account of improper planning of procurement which had bearing on transportation cost, input package and the disparity between lean and flush season supplies.

Studies on Cost Structure and Returns in Dairying

Studies on the structure and determinants of costs and returns are presented households very briefly.

Pillai[61] considered the cost of fodder feed, labour (both family and hired) veterinary and miscellaneous expenses as variable cost, and depreciation and interest on the amount invested in animals, cow-shed and dead stock as fixed cost. Among the fixed cost items shed and cows were significant, while feed ranks high in the variable cost.

Ramasubban and Goel[62] carried out a study on the cost of milk production in Delhi. The authors opined that the cost of milk production had declined with the increase in the level of milk production. The study clearly pointed out that the productivity of milch animals had gone up due to the implementation of Intensive Cattle Development Scheme.

Panes et al.[63] estimated the cost of production of milk for Tharaparkar, Sailiwol and Red Sindhi cows at National Dairy Research Institute. It was observed that the feed cost accounted for 60-70 per cent of gross cost. Further, per litre net cost and net income were Rs. 1.53 and Rs. 0.30 respectively.

Jodha and Choudary[64] found that, the cost of milk production decreased with the increase in herd size. They concluded that the seasonal variations affected dairy industry. Dairying became a profitable enterprise during monsoon and winter, while a liability in summer. The authors concluded that the organized marketing had a positive impact on the milk yield.

Parkala, et. al.,[65] analyzed the profitability of newly established dairy project at Mahatma Phule. It was estimated that the shares of working cost and fixed cost in the total cost were 73.38 per cent and 26.62 per cent respectively. Among the component of cost feed cost ranked first followed by labour and interest.

In his study *Venkata Krishnan,*[66] classified the milk processing cost into short and long-run costs, it was pointed that the long-run costs had relationship with the size of the plant, while short-run cost with the volume of milk produced.

Bhasin[67] pointed out that share of the feed in there of in the total cost of production was in the range of 60-70 per cent. In western countries it accounted for 45-60 per cent. The difference was due to the cost of labor.

Bhattacharya[68] investigating into the kind of impact on returns, found that of the average net return per Rs.100 invested was Rs. 16.60 in dairy farming unit, whereas Rs.13.80 in the arable farming unit.

Sambasiva Rao[69], analyzed the costs and returns of dairy farms so as to estimate the profitability of enterprises in different size groups of farms. The study concluded that the big and large farmers were in advantageous position in dairying Small and managerial farmers were unable to carry out dairy farming on profitable lines due to uneconomic landholdings, inaccessibility to production inputs and lack of extension education.

In his study *Jayachandra*[70] worked out season-wise cost and returns from dairying and employment generation in Chittor district. The author desired the increasing the green fodder supply so as to reduced the maintenance cost of milk animals.

Ram[71] studied the organizational and managerial aspects of dairy co-operatives in Rajasthan over a period of 5 years from 1979. The quantitative aspects of growth were analyzed, such as capacity in the processing the milk and manufacture of milk products. The input supply by the district union had remained inadequate and inconsistent. Finally, the author concluded that personnel policies were developed systematically and union management relationship was unhealthy.

Madan Mohan[72] empirically analyzed the management and functioning of dairy units in Warangal district. The author opined that the funds spent on the development of infrastructure and extension activity should not be considered as a liability on dairy unit. The author suggested the democratization of producers' cooperatives at the village level and provision of funds to sustain the extension activity.

Krishnaraj and Dubey[73] evaluated the external factors that affecting the organizational efficiency of milk producers cooperative societies. In their opinion organizational efficiency significantly depended on the size of milk producers cooperative society and supply of milk.

Ajit Kanitkar and Balavadi[74] examined the functioning of the Board of Directors (BOD) by analyzing the agenda items of board meetings of 13 cooperative milk unions. The authors concluded that the BOD is more concerned with the day to day operations than strategic issues.

PART-B: OBJECTIVES AND RESEARCH METHODOLOGY OF THE STUDY

Keeping the literature review in back ground, the research problem is stated, objectives are set and methodology to achieve the objectives in detailed.

Statement of the Problem

Milk is a perishable commodity and its surplus cannot be stored for a long time. Unlike agricultural produce, milk has short life; in most cases it is not more than three hours. Therefore, the farmer has to market produce within three hours of production twice a day and all the days of the year. The farmers who produce milk are at the mercy of milk vendors and the milk producers are often exploited. Another problem in milk marketing inadequacy of proper transport facilities, long distances to the nearest consumption centre, lack of proper roads, and means of transport and inability of the farmers to pool their milk and to organize collective methods of transport make it extremely difficult for the farmer to take his dairy produce to the market. While the farmers convert their milk into products, the methods of manufacture are outmoded and wasteful, the products are non-standardized and unhygienic and marketing arrangements are haphazard and unscientific. All these factors have resulted in on the one hand, in high price of milk and milk products to the consumers and, on the other hand, financial return to the producers well below the cost of production of milk. It would be possible to pay the farmers a higher price for their milk than what they obtain at present, if processing and marketing is done in systematic manner by reducing the overhead costs and improving the quality of marketing services. This would encourage the farmers to pay more attention to milk production. A beginning was been made in organized dairying in Kaira district with establishment of milk producers' co-operative societies at village level. Later, this scheme was been extended to the whole of the country under the Operation Flood Programme by the NDDB. Under this programme, the district level co-operative milk producers' union limited is expected to perform the functions of procuring the excess milk from the societies and sell them in urban areas, producing milk products such as ghee, butter, butter-milk, flavored-milk, pannier and kova (a milk sweet) from the excess of milk available, purchasing manufacturing and distribution of cattle feed, dairy machinery to the member societies, making prompt payment of money to the societies for the milk procured and supervising the working of the societies. Hence, an attempt is made to assess the performance of marketing products of dairy units.

NEED FOR AND SIGNIFICANCE OF THE STUDY

In view of the importance of dairying in Indian economy and the pivotal role of dairy products in the consumption basket of people, particularly in a developing economy like India, there is need to study the development of dairy industry at all levels. The need is felt to a greater extent from the research gaps identified in the light of the survey of relevant literature relating to dairy development in India. The brief review indicated that even though there are number of studies on economic aspects of dairy industry, the studies relating to marketing aspects have been relatively neglected. In the new era of liberalization and privatization when there is a surge of private dairy farms, there is cut throat competition in the market. Hence, marketing of dairy products must be given due importance. Therefore, there is an urgent need to appraise the marketing activities of dairy industry. The fact that there are only quite a few studies on performance of dairy industry in Andhra Pradesh in general and marketing aspects in particular also strengthens the case for a study on marketing practices of dairy units in Kurnool district of Andhra Pradesh.

In a district, the marketing of dairy products involves three types of agencies. There are Co-operative Milk producers Union at the district level, Milk Producers Co-operative Societies and the individual milk producers at the village level. The Union, the societies and the members involved in the process face different problems in marketing of milk and milk products at the levels of production, procurement, and the fixation of sale prices of milk and milk products. These problems stand in the way of dairy development. Hence, it is necessary to study different aspects of marketing and the problems involved in marketing of dairy products. Accordingly, an attempt is made here to study the marketing of dairy products of Kurnool District Co-operative Milk Producers Union, Kurnool, hereafter referred to as KDCMPU. This study would highlight the marketing practices of the dairy units of Kurnool district and the suggestions of this study will be helpful to the managements to take necessary steps to remove the present defects in functioning of their dairies.

OBJECTIVES OF THE STUDY

Following are the objectives set in the study:

(*i*) To review the progress of dairy industry in India and Andhra Pradesh;

(*ii*) To examine the trends in milk procurement by sample dairy units;

(*iii*) To set forth the product-related marketing practices of sample units;

(*iv*) To analyze the pricing objectives and practices of select units;

(*v*) To assess the promotions mix of the dairy units; and

(*vi*) To evaluate the sales and distribution practices of dairy units.

Sample Design

The main focus of the study is to evaluate the marketing practices of dairy units in Kurnool District of Andhra Pradesh. Therefore, the universe is restricted to Kurnool district only. There are six dairies in Kurnool district; of them four undertake only procurement of milk. Therefore it is inevitable to select the remaining two dairies undertaking all aspects of marketing of dairy products. These dairies are Vijaya Co-operative Dairy located at Nandyal, and the other, Nandi dairy in private sector also located in Nandyal. Therefore, the researcher has to include the aforesaid dairies in to the sample frame.

Data Sources

The secondary data pertaining to 10-year period, 1998-2007 were collected from the Vijaya Co-operative Milk Producers Union and Nandi Dairy Pvt. Ltd. These data from include internal records and audited annual reports. Apart from these internally available data, secondary data have been compiled from the magazines, journals, dissertations, government statistical handbooks, *Dairy India Year Book* and other publications from various institutions and other sources in order to have a micro and data macro-levels of websites are also installed for necessary data.

Tools of Analysis

The data, thus, collected are tabulated, analyzed and interpreted with the help of appropriate statistical tools and techniques. These include percentages, annual growth rates, averages, standard deviations coefficient variation, simple linear regression, t-test. Besides these tools, diagrams, graphs and charts are used graphically to highlight presentation of numerical data in tabular form.

Scope and Limitations

The study addresses issues faced in marketing by dairy units in Kurnool district. Kurnool district has been purposively selected the area operation of the two major dairies in Kurnool district i.e. Vijaya Co-operative Milk Producers Union Limited and another one is Nandi dairy in this study focus on procurement of milk and production of milk products, dairy units communication mix pricing practices and distribution of milk products are analyzed in details in the study which really determine the surviving capacity of the dairy industry. Thus the result obtained from the empirical study can be used for generalization in policy decisions in other areas of the country. However, the study is not without limitations. It is two dairy units one is private sector and other in co-operative sector, operating in the Kurnool district. The conclusions of the study cannot be generalized area. It is the study confined to secondary data focusing on marketing practices adopted by sample units, learning out the consumers milk producers and dealers.

Organisation of the Book

This book is organized into nine well demarcated chapters. Chapter one is an introductory in which the development of dairy industry is shown at the national and State levels. Chapter two are presented select review of literature and research methodology. Profiles of sample dairy units are graphically illustrated in chapter three. Trends in milk procurement form part of chapter four. Chapter five sets forth product-related marketing practices of sample dairy units. Pricing objectives and methods are dealt in chapter six. Promotion mix of dairy units is the subject matter of chapter seven. Practices in the domain of sales and distribution are described in chapter eight. Chapter nine is devoted to findings, conclusions, suggestions and areas for future research.

REFERENCES

1. Girdhari, D.G, "Dairy Marketing", *Indian Journal of Marketing,* Vol. XII, No. 11, 1989.
2. David, Avery Vose, "Market Structure, Conduct and Performance at the Midwest Dairy Industries", Unpublished Ph.D thesis, Madison University at Wisconsin, 1966, p. 17.

3. Shaik, N.A., "Liquid Milk Marketing", *Indian Dairy*, Vol. 40, No. 5, 1988, pp. 291-292.
4. Kaur, Mainder and Gill, G.S, "An Economic Analysis of Marketing of Milk in Ludhiana District (Punjab)", *Agricultural Marketing*, Vol,XXXI,No.4, January-March 1989, pp. 19-21.
5. Sukumar, D.E., *Outlines of Dairy Technology*, Oxford University Press, Bombay 1980, pp. 1-3.
6. Kalra, K.K. and Raj Vir Singh," Optimizing Milk Marketing System in Organised Sector, *Indian Journal of Agricultural Economics*, Vol. XXXIV, No. 4, 1984, pp. 208-211
7. Patel, A.S., *Co-operative Dairying and Rural Development: A Case Study of Amul,* Oxford University Press, New Delhi, 1988, pp. 370-376.
8. Ram, Jawana, Marketing of Milk and Milk Products: A Case Study of RCDF, *Indian Co-operative Review*, Vol. XXVIII, January 1991, pp. 225-226.
9. Chahal, S.S. and Balwindersing, An Economic Analysis of Activities of MILKFED in Production and Marketing of Milk, *Indian Co-operative Review*, Vol. XXX, No. 4, 1993, pp. 360-362.
10. Mandana, P.K. and Srinivasagowda, M.V, Co-operatives and the Commercialization of Milk Production in India, *Southern Economist*, Vol. 33, No. 14, 1994, pp. 13-14.
11. Sah, A.K., Dairy Scenario: Rising Hope, Yojana, Nov. 1998, Vol.42, No. 11, pp. 42-44.
12. Prabharan, R., Research Investment Crucial, Survey of Indian Agriculture, *The Hindu*, 2000, pp. 137-140.
13. Murthi, S. Suriya, 2001. Milk Marketing Strategies, *Indian Journal of Marketing*, Vol. 31, No. 56, pp. 24-25.
14. Kumar, S.M., 2002. "Procurement and Distribution of Milk: A Case Study with Reference to Villupuram District Cooperative Milk Producers' Union Limited", Unpublished M.Phil Thesis, Annamalai University.
15. Pitchai, C., 1999. A Study on the Distribution of Milk and Milk Products by Co-operatives in Tamil Nadu, Unpublished Ph.D Thesis Annamalai University.
16. Chahal, S., 1998. "Performance of Co-operatives in Marketing of Milk in Punjab", *Indian Co-operative Review*, p. 274.
17. Sanghu, K.P.S., 1994. "Production, Consumption and Marketed Surplus of Milk in Western Uttarpradesh", *Indian Dairyman*, Vol.46, No. 10, p. 629.

18. Jairath, 1981. Marketing of Milk: A Study of Price Behaviour, *Indian Journal of Marketing*, Vol. 11, No. 9.

19. Murali Manohar, D. and G.S. Sundharshan, 1982. "Dairying as A Household Industry", *Kurukshethra*, Vol. 30, No.11, p. 10.

20. Patel, R.K., and R.Prabaharan. 1980. "Consumer Awareness and Preference for Milk in Madras City", *Indian Journal of Marketing* Vol. 11, No. 4, pp. 13-16.

21. Chinnaiyan, P., Kanadasamy and G. Rangarajan, 1980. "Marketing of Milk in Erode Taluk, Tamil Nadu", *Indian Journal of Marketing*, Vol. 10, No. 8, pp. 21-26.

22. Ravikiran, G.N., V.Jayarama Krishnan, and A.Salhynarayana, 1994. Situtional Analysis of Milk Production in Krishna District Milk Shed of Andhra Pradesh, *Indian Dairyman*, Vol. 46, No. 12, p. 741.

23. Gupta and Devarajan, 1995. Consumption and Disposal of Milk in Churu District (Rajasthan), *Indian Dairyman*, Vol. 47, No. 6, pp. 42-45.

24. Patel, S.M., and M.K. Pandey. "Economic Impact of Kaira District Co-operative Milk Producers Union (Amul Dairy) in Rural Areas of Kaira District". Institute of Co-operative Management, Ahmedabad, 1976, p. 8.

25. Myrdal, Gunnar, *Asian Drama*, New York, Random House, 1968, pp. 260-61.

26. Mishra, S.N., *Livestock Planning in India*, New Delhi, Vikas Publications, 1978.

27. Ramachanadran, L, "Indian Food Problems", A New Approach, Bombay, Allied Publishers, 1977.

28. Grill, S.S. and Grill, G.S., "An Analysis of Economic and Social Change Through Dairying: Dairying as an Instrument of change", India, XIXth *International Dairy Congress*, 1974, pp. 83-89.

29. Suhag, L.R, Vijay Kumar and Anil kumar Rabhee, "Role of Dairy Animals in Income and Employment Generation of Different Sizes of Farms in Haryana", *Indian Dairyman*, Vol. XVII, No. 7, July 1985, pp. 295-299.

30. Meliville, A.R, *The Task of Dairy Development Dairying is an Instrument of Change,* New Delhi, Mittal Publications, 1974.

31. Achrya, T.K.T. and Pawar, T.G. "The Economics Production by Different Types of Milch Animals", *Indian Journal of Agricutural Economics*, Vol. 35, No. 4, 1980, p. 153.

32. Ranga Rajan G, Ramaswamy and Puhayhandi, "Milk Production by Different Types of Milch Animals", Indian Journal of Agricultural Economics, Vol. 35, No. 4, 1980, p. 153.

33. Sharbuddin, "Impact of Dairy Cooperatives on the Economy of Farmers", Ph.D. Thesis, , S.K.University, Anantapur, Andhra Pradesh 1986.

34. Bhanja S.K and Venkatadri S, "Milch Cattle in IRDP", Hyderabad, National Institute of Rural Development, 1988.

35. Biradar, R.D, "Rural Development Through Dairy Development: A Case Study", Kurushetra, Vol XXXVI, No. 5, Februry 1988, pp 3-6.

36. Krishna Raju, G. and Seshaiah, K. "Dairy Development in India – An Antimical Study", Khadi Gramodyog, Vol. No. 7, March 1990, pp. 268-277.

37. Mitra, M. "Profiles of Women Dairy Producers in A.P", Operation Flood and Indian Dairying, New Delhi, Sage, 1990, p. 300-315.

38. Somjee, A.H. and Geeta Somjee," Dairy cooperatives: A Catalyst for Economic and Social Change in Rural India", Indian Dairyman, Vol. XXII, No. 8, Aug. 1990, pp. 341-347.

39. "Operation Flood Benefits Rural Poor", NCAER Indian Dairyman, Vol. XXIII, No. 4, April 1991, pp. 201-202.

40. Sekhar Dutt, "Dairy as an Investment of Development", Yojana, Vol. 34, No. 9, May 1990, pp. 18-20.

41. Thirunavukkarasu, M. Prabhakaran,P. and Ramaswamy,C, "Impact of Operation Flood on the Income and Employment of Rural Poor: Some Micro Level Studies", Journal of Rural Development, Vol. No. 2, July 1991, pp. 37-45.

42. Somjee, A.H., and Somjhee, G., Managerial Shift, World Animal Review. Vol, XXIV, No.11, 1976, pp. 28-33.

43. Mitra, M., Womens Work: Gains Analysis of Womens Labour in Dairy Production, Women and the House Household in Asia-1, Sage Publication, New Delhi, 1987, pp. 109-111.

44. Nageswara Rao, S.B., All Women Milk Cooperatives in A P. Kurukshetra, Vol XXXV, No, 11, August 1987, pp. 29-31.

45. Mascarenhas, A Strategy for Rural Development, and Dairy Cooperatives in India, Sage Publications, Bomabay, 1988, pp. 295-297.

46. Miriam Sharma and Urmila Vanjani, Womens Work in Never Done Dairy Development and Health in Lives of Rural Women in Rajasthan, Economic and Political Weekly, Vol. XXIV, No. 17, 1987 pp. 38-43.

47. Kulandaisamy, V. Women's Participation in Dairy Cooperatives, *The Tamilnadu Journal of Cooperation*, Vol. 83, No. 1, 1991, pp. 23-26.

48. Subramanyam, Usha, Women's Dairy Cooperatives, *Kisan World* Vol. 56, No.13, 1996, pp. 52-53.

49. Mudgal, V.D., Role of Livestock in Indian Farming System: An Analysis, *Indian Dairyman*, Vol. 51, No. 5, 1999, pp. 25-27.

50. Gangwar A.C, "Inter District Variation in Bovine Population Cost Structure, Production and Disposal of Milk in Haryana", *Indian Journal of Economics* Vol. 30, 1975, p. 140.

51. Kumar P. Patel. R.K and Raut K.C, "Lactation-wise Production Function and Concentration in Milk Production on Haryana Cows," *Indian Journal of Agricultural Economics*, Vol. XXX, No. 37, 1975, pp. 128-133.

52. Shah, Deepak and Sharma, "Supply Functions for Milk in Buludhser District of U.P", *Indian Journal of Agricultural Economics*, Vol. 59, No. 10, 1995, pp. 745-749.

53. Gupta, Raghuraj, "Optimising Milk Production," *Indian Journal of Rural Development*, Vol. XXIII, No. 11, 1975, p. 13.

54. Sharma, K.N.S., Patel, R.K. and Surendra Singh, "Projections of Bovine Population, Requirement and Economic Demand for Milk", *Indian journal of Agricultural Economics*, Vol. XXX No. 3, 1975, p. 142.

55. Gupta, V.K and Pandey, R.K, "Consumption and Availability of Milk in India", *Indian journal of Agricultural Economics*, Vol. XXX No. 3, 142, 1975.

56. Shanti George. "Diffusing Anand: Implications of Establishing a Dairy Cooperative in a Village Central Kerala, "Economic and Political Weekly, Vol. 19, No. 51, December, 1984, pp. 2161-2169.

57. Kumar, P. and Ranf, K.C. "Some Factors Influencing the Economy of Milk Production", *Indian Journal of Agricultural Economics*, Vol. XXVI (2), 1971, pp. 129-136.

58. Sankayam and Joshi, "Resource Productivity in Milk Production of Crossbred and Indigenous Cows in Rural Areas of Ludhiana District, *Indian Journal of Agricultural Economics*, Vol. XXX (3), 1975, pp. 105-106.

59. Muranjan, S "Factors Responsible for Increased Procurement of Milk in Maharastra, *Artha Vijnana*, Vol. 29, No. 4, December, 1977.

60. Madhavan, E "Procurement Planning and Scheduling", *Indian Dairyman,* Vol. XXX, No. 5, 1978, pp. 303-308.

61. Pillai, Maniekya Vasagam, N, "A study on Resource Use Efficiency in Milk Production in Parambikulam Aliyar Project Region", Tamil Nadu, Dissertation Submitted to the Tamil Nadu Agricultural University, Coimbatore, 1976.

62. Ramsubban, T.A. and Goel S.K, "An Enquiry in to the Cost Consumption and Supply of Aspects of Milk Produces by Cultivating Families in Delhi Area", *Indian Journal of Agricultural Economics*, Vol. XX, No.1,January-March,1965, pp. 92-97.

63. Panse, V.G., *et al.,* "Cost of Milk Production in Madras", ICAR Report Series No. 10, 1963.

64. Jodha, N.S.and Chourary K.M," Prospects and Problems of Dairy Development in a Desert Region", Agro-Economic Research Centre, Vallabh vidyanagar, 1970(Mimco).

65. Parkala, D.G.,Kasar D.V and D.R.Pise, "An Investigation in to the Economics of Milk Production on a Newly Establish Dairy Farm", *Indian Journal of Agricultural Economics*, Vol. XXX No. 3,1975, pp. 2-43.

66. Venkatakrishnan, "Method of Estimating Costs in Dairy Plant", *Indian Dairyman*, Vol.XXVII, No. 8, 1975, p. 295.

67. Bhasin, M.R, "Exploring Ways and Means to Reduce the Cost of Production of Milk to Encourage Dairy Farming", *Indian Dairyman*, Vol. XXVII (7), 1975, p. 273.

68. Bhattacharya, "New Strategy in Animal Husbandry", *Indian Dairyman*, Vol. XXXIII, No. 2, 1976, p. 54.

69. Sambasiva Rao B. "Economics of Dairy Farming: A study", *Indian Dairyman*, Vol. 38, No. 3, 1986, pp. 105-109.

70. Jayachandra, K, "Dairying in Drought Prone Areas: A study", *Yojana*, Vol.34, No. 4, March, 1990, pp. 27-29.

71. Jawana, Ram, "Management of Dairy Enterprises", Jaipur: Kuber Associates and Publishers, 1987.

72. Madan Mohan C, *Dairy Management in India: A study in Andhra Pradesh*, New Delhi, Mittal Publications, 1989.

73. Krishnaraj R and dubey V.K. "External Factors Affecting Organisational Efficiancy of the Milk Producers' Cooperative Society", *Indian Cooperative Review*, Vol XXXVII, No. 4, April 1990, pp. 345-351.

74. Kanitkar, Ajit and Belavadi, N.V, "Board of Directors: Theory and Practice in Dairy Co-operatives", *Vaikalpa*, Vol. 18, No. 4, October-December 1993, pp. 21-27.

3 CHAPTER

PROFILES OF SAMPLE DAIRIES

INTRODUCTION

This chapter presents a brief account of Kurnool district such as its geographical location, climate and revenue set-up. Besides this, it also discuses, in two separates parts, the profiles of sample dairies located in Kurnool district.

Geographical Location of Kurnool District

Kurnool district was named about its chief town Kurnool, from first November 1956. The name 'Kurnool' is said to have been derived from 'Kandanavolu'. The geographical location district lies between the north latitude of 14° 54' & 16° 18' and east latitude of 76° 58' and 79° 34'. The altitude of the district is above 100 ft above the sea level. The boundaries are district are Tungabhadra and Krishna rivers as well as Mahaboobnagar district on the north, Kadapa and Anatapur districts on the south, Bellary district of Karnataka state on the west, and Prakasam district on the East.

Climate Conditions and Revenue Administration Set-up

The climate of the district is normally good and healthy usually months of January, February and March are pleasant with moderate winds from Southeast. April and May are the hottest months of the year. During these months, winds shift to South-east with increased force and bring welcome showers by the end of May. During the succeeding four months, the winds blow from western side in the major part of the district and bring fair quantum of rain fall. By the end of September, the wind is light and pleasant, forecasting the on-set of North-east monsoon. During November and December, the weather is fine. Rainfall is rare and wind is light with occurrence of heavy dew. During 2005 the district received 584.2mm rain fall as against normal rainfall of 670 mm. There are three revenue divisions, 54 revenue mandals, 53 mandal praja parishads, one corporation, 4 municipalities, 898 gram panchayats, 920 revenue villages and 615 hamlets.

Infrastructure for Dairy Development

The climate conditions in the district are highly conducive for developing dairy industry. The demand for milk far exceeds supply. The milk producers' union buys milk from neighboring districts. To increase milk supply, the state government as well as producers union has been encouraging the villagers to start milk co-operative societies. Further, under IRDP, the central government compelled the commercial banks provide loans and advances liberally to villagers, especially down-trodden farmers. The village milk co-operative societies provide fodder and veterinary services to their members. Thus, dairy gained momentum in the district next to agriculture. The livelihood of poor villagers depends upon on the successful functioning of village milk co-operative societies. These societies encourage active participation of women in dairying, thereby enhancing their status in their families.

PART-A: PROFILE OF VIJAYA DAIRY

Origin of KDMPMACU

APDDC started its unit with the capacity of 12500 lts per day, at Kurnool during 1974. Later on during 1977, the APDDC established a chilling centre at Nandyal with capacity of 30000 lts per day. The APDDC started the Kurnool District Milk Producers Co-operative Union Ltd. (KDMPCU) at Kurnool, and the chilling centre factory at Nandyal was expanded with a capacity of 1.5 lakh lts per day. This was possible with financial support 10 crores made available by the NDDB. The KDCMPU was converted into Kurnool District Milk Producers Mutually Aided Cooperative Union Ltd. (KDMPMACU). This dairy was named Vijaya Dairy and its brand is Vijaya is well known to all. This dairy is supported by two major chilling centres located at Mydukur and Chagalamarri and twelve bulk cooling /chilling centres located at twelve places such as Dhone, Banavasi, Pathikonda, Halaharvi, Atmakur, Kolimiguldla, Rayachoti, Pulivendula, Dornipadu, Sankalapuram, Proddutur and Rajampeta. These centres are spread over Kurnool as well as Kadapa district. Presently KDMPMACU Ltd., produce' one lakh litres milk per day.

There are 312 employees in its organisation. It procures quality milk, brings out quality milk and milk products, supplies 10 thousand lts of milk per day. Due to the fact that the milk is of supervisor quality market for *ghee* is spread over five States like Maharashtra, Karnataka, Tamil Nadu, Rajasthan and Orissa, besides Andhra Pradesh (AP). The market for this dairy is vast, compared to Nandi dairy due to manufacture of quality products, organisational and marketing network.

Services of KDMPMACU Ltd.

Veterinary first aid and insemination centres well-equipped and manned by well-trained personnel attend first aid cases and involved are in vaccination. With seven surgeons of KDMPMACU Ltd., provides Veterinary services, Medicines, tinctures and powder. The union further supplies liquid nitrogen to the Artificial Insemination Centres (AICs) which is distributed through its vehicles. The union bears the total expenditure incurred in this regard. Three months after artificial insemination, pregnancy will be verified. Procurement route Vete: doctors regularly follow up 'calf' till 10 months from calf's birth. They also recommend incentives to AI workers.

Fertility Campaigns

The union conducts fertility campaigns at dairy cooperative societies, covering over 50,000 animals every year through fertility campaigns. The route doctors look after fertility cases in the diagnosis, treatment and follow-up stages.

Vaccination

The union organisers' vaccination programmes, bearing 50 per cent cost the rest bearing milk producers under vaccination programme. The dairy provides foot and mouth disease vaccinations to nearly 35000 animals HS around 5 thousand animals, anti-rabies to about 50 animals. The vaccinations lead to long life and health to animals, thereby augmenting milk from animals.

Provision of De-worming Drugs

The company supplies de-warming drugs to societies free of cost to save the lives of calves, calf the future milk animals. These drugs kill parasites in this stomach and maintain the calf/animal good health. This drug saves the calf from life threat and increases milk yield from animals. During 2007 the company made available 36000 doses of de-worming drugs to milk producers.

Insurance

In the union initiated insurance for milk animals half of the insurance premium is forge by the union and the remaining half by the farmer. The union has been executing this programme since 2005, in the during year 2,760 animals were covered under this programme.

Feed

The KDMPMACU Ltd. provides concentrated feed to improve health of milk animals and thereby to extent more milk. Every month forty farmers require cattle feed. During 2007 the union distributed 460 tonnes of feed and 50 tonnes of mineral mixture.

Urea Treatment

The union implements a new programme with a view to enrich the nutritive value of paddy and jower straws through urea treatment process. This process is economical due to fact that the cost of production of milk substantially declines.

Others Services

The dairy distributes sophisticated electronic instrument to cooperative societies so as to correctly measure the volume and the fat content in the milk. Consequently farmers are paid fairly. In other words it eliminates the under weighment and arrives at correct fat content. The union acquired 60 cm projector and TV with VCP. During evenings films relating to cattle health, milk production and benefits of becoming members of cooperative societies would be exhibited to the producers. To create awareness among the farming community, KDMPMACUL. users different kinds of media. Advertisements regarding dairy products are printed on the milk sachets. Further literates are enlightened by media advertisement in news papers, and periodicals and booklets and pamphlets. Hoardings are also prepared and exhibited to promote sales of milk and milk products.

Milk Procurement and Prices are Discussed in this Section

As can be seen from Table 3.1 KDMPMACU Ltd., procure milk from 36000 milk producers spread over 566 villagers in Kurnool and kadapa district's, there are 19 procurement routes. The company procures milk through MPCS/MPAC milk Chilling centres and bulk units. There are 256 MPCS, while MPACs are 463. The company engages 35 own/rented vehicles to transport milk from procurement centres/ cooperatives, and chilling centres.

Reading Table 3.2, it can be observed that the daily and annual procurement of milk during 1998-2007. During 2007 on an average per day, 88268.49 litres were procured and as against 25227.29 litres during 1998. It may be observed that

Table 3.1. Select Particulars of Vijaya Dairy as on 31-12-2007

Sl. No.	Particulars	Nos.
1	No. of MPCS	256
2	No. of MPACs	463
3	No. of Procurement Routes	19
4	No. of Villages Covered	566
5	Milk Producers	36000
6	No. of Milk Chilling Centres	3
7	Transport Vehicles Own/Rented	35

Source: Compiled from the records of Vijaya dairy.

Table 3.2. Daily and Annual Procurement of Milk: From 1998-2007

Sl. No.	Year	Daily average milk procurement (in litres)	Annual procurement k(in lakh litres)
1	1998	25227.39 (....)	92.08
2	1999	30224.65 (19.80)	110.32
3	2000	44706.84 (47.91)	163.18
4	2001	56339.72 (26.02)	205.64
5	2002	89331.50 (58.55)	326.06
6	2003	68241.09 (-23.60)	249.08
7	2004	79986.30 (17.21)	291.95
8	2005	94665.75 (18.35)	345.53
9	2006	81649.31 (-13.75)	298.02
10	2007	88268.49 (8.10)	322.18

Note: Figures in parentheses are annual percentages increases ones the preceding years.

Source: Compiled from the records of Vijaya dairy.

highest quantum of milk 94665.75 litres was acquired in 2005, there are ups and downs in the intermittent period. The growth in 2007 over 1998 is 3.49 times. Annual percentage increase in average daily milk procurement was highest at 58.55 in 2002, and lowest at 23.60 in 2003. During 1998 in 92.08 lakh litres were procured as compared to 322.18 lakh litres in the terminal year 2007, registering more than three-fold increase. From the preceeding analysis it can be concluded that procurement by the daily, on daily average basis as well as year period has kept increasing year after year, exceptions being procurement in 2003 and 2006.

The prices offered to milk producers by Vijay Dairy per litre of milk corresponding to fat percentage levels during 2007 are given in Table 3.3. The lowest price offered is Rs. 12 per litre for a fat content. For every 0.1 per cent increase in fat content, the price offered increases by Rs. 0.24 per litre. For a maximum of 10 per cent fat, the price offered for litre is Rs.24.00. Said in other words the price for litre doubles as the fat per cent increases twice.

Further the aforesaid prices are subject SNF per cent. If the SNF is 8.8 per cent are more the aforesaid prices are paid. From the above analysis it can be concluded that there is a price per litre which is based on both fat and snf fixes percentages. It means that the company follows uniform prices policy in processing milk from the formers.

In Table 3.4 are shown price per litre sliding down as the SNF percentage declines. To be specific, 0.1 per cent decline in SNF results in 6 paise reduction in price per litre. SNF percentage sliding down to the maximum level of 8.8 to 8.0 attracts price reduction from 'o' paise to 48 paise per litre.

Contents and Shelf Life of Milk and Milk Products

A comparative analysis of a cow and buffalo milk is reported in Table 3.4. The buffalo milk compares favourably with cow milk in respect of fat and protein contents.

Table 3.3. Milk Procurement Prices *vis-à-vis* Fat Percentages in Vijaya Dairy During 2007

Fat %	Milk price per/ litre	Fat %	Milk price per /litre	Fat %	Milk price per/litre
5.0	12.00	6.7	16.08	8.4	20.16
5.1	12.24	6.8	16.32	8.5	20.40
5.2	12.48	6.9	16.56	8.6	20.64
5.3	12.72	7.0	16.80	8.7	20.88
5.4	12.96	7.1	17.04	8.8	21.12
5.5	13.20	7.2	17.28	8.9	21.36
5.6	13.44	7.3	17.52	9.0	21.60
5.7	13.68	7.4	17.76	9.1	21.84
5.8	13.92	7.5	18.00	9.2	22.08
5.9	14.16	7.6	18.24	9.3	22.32
6.0	14.40	7.7	18.48	9.4	22.56
6.1	14.64	7.8	18.72	9.5	22.80
6.2	14.88	7.9	18.96	9.6	23.04
6.3	15.12	8.0	19.20	9.7	23.28
6.4	15.36	8.1	19.44	9.8	23.52
6.5	15.60	8.2	19.66	9.9	23.76
6.6	15.84	8.3	19.92	10.0	24.00
X				8.00	15.60

X of FAT % age = 8%
X of price per ltr of milk = Rs 15.60.
Source: Compiled from the records of Vijaya Dairy.

The fat parentage is 7 in the buffalo milk and 3.5 per cent in cow milk. Similarly, proteins constitute 5 per cent and 4.6 per cent in the former and latter respectively. A converse situations emerges in respect of lactose, ash and water contents as they constitute five per cent 0.8 per cent and 87

Table 3.4. Reduction in Milk Price per Litre Corresponding to Reduction in SNF in Vijaya Dairy during 2007

Sl. No.	Fat (%)	Milk price (Rs. paise)
1	8.8	Nil
2	8.7	0.06
3	8.6	0.12
4	8.5	0.18
5	8.4	0.24
6	8.3	0.30
7	8.2	0.36
8	8.1	0.42
9	8.0	0.48

Source: Compiled from the records of Vijaya Dairy

Table 3.5. Content of Buffalo and Cow milk in Vijaya dairy

Sl. No.	Milk Contents	Buffalo Milk	Cow Milk
1	Fat	7%	3.5%
2	Proteins	5%	4.6%
3	Lactose	4.2%	5%
4	Ash	0.7%	0.8%
5	Water	83%	87%

Source: Compiled from the records of Vijaya Dairy

per cent in the cow milk whilest 4.2 per cent and 0.7 and 83 per cent in buffalo milk. It can be concluded that the buffalo milk in certain aspects and cow milk contain in other aspects are found relatively superiors.

Price Fat and SNF Contents and Shelf Life of Milk Products and Milk by Products of Vijaya Dairy

Table 3.6 presents five kinds of milk products of Vijaya Dairy and their contents features. These products carry Agmark quality of agricultural/dairy products attestation by the Govt. of India. Further ISI mark denotes the meeting of standards set by the Bureau of Indian Standards (BIS). 'The Vijaya' brand name stands for purity and quality as a result of millions of consumers all over the country trusting it. The fat and snf percentages vary from product to product. In gold milk, fat content is 6 per cent while SNF content is nine per cent. This milk is packed in with red labels. With regard to toned milk former and later constituted three per cent and 8.5 per cent respectively. It is available in green labels with 500 ml and 200 ml packets. The fat content accounts for 1.5 per cent and SNF nine per cent in double toned milk which is packed in brown. In standardised milk the former and later constitute 4.5 per cent and 8.7 per cent respectively. The percentage of fat and snf in skimmed milk is 0.1 per cent and 7 per cent.

Table 3.6. Price Fat and Snf Contents and Shelf life of Vijaya Dairy Products

Sl. No.	Milk products	Price	Fat%	SNF%	Shelf life
1	Gold milk	20.00	6	9	One day at Shelf life
2	Toned milk	18.00	3	8.5	One day at Shelf life
3	Doubled toned milk	17.00	1.5	9	One day at Shelf life
4	Standardized milk	15.00	4.5	8.7	One day at Shelf life
5	Skimmed milk	12.00	0.1	8.7	One day at Shelf life

Source: Compiled from the records of Vijaya Dairy.

As can be deserved from Table 3.7 is Vijaya Dairy manufacturers pannier, basic product, and by-products such as butter, *ghee*, skimmed milk powder, *doodhpeda*, butter milk and sterilized flavoured milk. The fat content in paneer is 99 per cent snf is nill. In by-product ghee has fact content 99 per cent and self life of six month. Butter has a shelf life of three months, fat per cent of 84 and SNF per cent 2, sterilized flavoured milk has 1.5 and 9 per cent fat and snf, contents shelf life of six months respectively and doodhpeda has shelf life of 5 days, the fat and SNF percentages are 24 and 35 per cent respectively. Butter milk has shelf life of one day, and 1.4 per cent and 4 per cent fat and snf contents respectively. The skimmed milk powders a life span of one year and contains one per cent fat and snf 96 percent snf. Pannier has 99 per cent fat and shelf life of three days. It can be summed up that Vijaya Dairy producers paneer as basic product and by-product in addition to milk with different fat and SNF percentages and shelf life.

Table 3.7. Price, FAT and SNF contents and Shelf life of By-products of Vijaya Dairy

Sl. No	Products	Price	Fat%	SNF%	Shelf Life
1	Ghee (1000g)	160.00	99.9	_.....	6 Months
2	Butter (1000g)	140.00	84.0	2	90 days
3	Sterilized flavoured milk (250ml)	10.00	1.5	9	180 days
4	Doodhpeda	120.00	24.0	35	5 Days
5	Butter milk (200ml)	4.00	1.4	4	1 Day at shelf life
6	Skimmed milk powder(SMP)	130.00	1.1	96	1 Year
7	Panneer (1000g)	160.00	99	.....	1 Month

Source: Compiled from the records of Vijaya Dairy.

Sales of Milk, Milk Products and By-products in Vijay Dairy

Table 3.8 furnishes the daily and annual sale of milk by Vijay Dairy for the reference period, from 1998 to 2007. The Vijay dairy increased its dairy sale of milk from 21265.75 liters in 1998 to 52863.01 litres in 2007, registering over 2.48 times increase. A similar trend in average annual sales is aforesaid. The dairy sold 192.95 lakh litres in 2007 as compared

Table 3.8. Daily and Annual Milk Sales of Vijaya Dairy: From1998 to 2007

Sl. No.	Year	Daily average milk sales (litres)	Annual milk sales (lakh litres)
1	1998	21265.75 (....)	77.62
2	1999	20224.65 (-4.90)	73.82
3	2000	20583.56 (1.77)	75.13
4	2001	20923.28 (1.65)	76.37
5	2002	49049.31 (134.42)	168.08
6	2003	25865.75 (-47.26)	94.41
7	2004	32150.68 (24.29)	117.35
8	2005	37789.04 (17.53)	137.93
9	2006	52893.15 (39.97)	193.06
10	2007	52863.01 (-0.06)	192.95

Source: Compiled from the records of Vijaya dairy

77.62 lakh litres in 1998. Meanwhile, there are wild fluctuations. For example 168.08 lakh litres were sold in 2002 as against 94.41 lakh litres in 2003. Focusing on yearly percentage changes in daily sales as shown in parentheses of column three, it can be found that percentage changes are negative in 3 out of 9 years and range from –0.06 per cent in 2007 to -47.26 per cent in 2003. In contrast the percentage changes in annual sales are positive in 6 out of years, and range from 1.77 per cent in 2000 to 134.42 per cent in 2002. It can be inferred that there has been ups and downs in growth during the study period.

Table 3.9. Sales of By-Products of Vijaya Dairy: From 1998-2007

Year	Butter (tonnes)	Ghee (tonnes)	Skimmed milk powder (tonnes)	Doodh peda (tonnes)	Butter Milk (pockets)	Sterilized flavoured Milk (number of bottles)	Paneer (kgs)
1998	406.32	355.60	141.78	10.40	175569	253265	.....
1999	496.09	353.70	138.04	9.30	156227	226890	......
2000	607.01	348.04	50.32	9.70	142507	218135	......
2001	293.25	289.50	78.05	10.10	217251	279151	484.50
2002	329.04	254.65	114.40	12.1	253484	238967	699.45
2003	370.20	152.40	235.43	13.70	309691	228470	1626.35
2004	417.45	228.65	260.35	12.84	22987	232720	2358.90
2005	334.25	201.25	289.35	15.75	319085	211040	5938.45
2006	94.05	196.25	250.00	17.54	515727	163920	6995.30
2007	424.65 (4.51)	210.56 (-40.78)	258.54 (82.35)	21.45 (106.25)	701429 (299.52)	191060 (24.56)	7225.40 (1391.31)

Note: Figures in Parentheses in the last row are percentages variations over the year 1998

Source: Compiled from the records of Vijaya Dairy.

As can be seen from Table 3.9 the dairy sold 406.32 tonnes of butter in 1998 as against 424.65 tonnes in 2007, registering 4.51 per cent increase over the initial year. Sales of butter were least 94.05 tonnes in 2000. Skimmed milk powder, *doodhpeda* and butter milk showed an increase in sales in 2007 over 1998, registering 82.35, 106.25 and 299.52 per cent respectively. With regard to ghee and sterilized flavoured milk a contrasting picture is present sales of ghee and sterilised flavoured milk were 210.56 tonnes and 191060 bottles in 2007 as against 355.06 tonnes and 253265 bottles in 1998, registering -40.78 and -24.56 per cent increases respectively. These percentages were sold for different prices based on fat content and snf percentage. The sale price of ghee per kg was Rs. 160 and of butter milk Rs. 4 for 200 ml in 2007. The sale price per kg of butter was Rs. 140, of skimmed milk powder Rs. 160 and doodhpeda Rs. 120. The sterilized flavoured milk with 200 ml was offered at Rs. 10. It can be concluded that there has been growth in the production as well as sales of butter, skimmed milk powder, doodhpeda, butter milk and pannir. Contra picture is attained with regard to in the case of butter and sterilised flavoured milk.

Organization chart depicting various departments and hierarchical positions in each of these departments are shown in Chart 3.1. At the top is board of directors headed by chairman, which is the top policy making body of Vijaya dairy. Managing Director, who is full-time top management functionary next to the board, is in the second layer. After managing directors are heads of functional departments. The functions are production, plant, quality control, marketing finance and administration, and public relations are shown. The hierarchical levels are variously designated such as deputy directors; deputy general managers, and managing officer. This is top-down chart showing line authority, that is, arrangement are managers firm top to bottom with clearly defined line of command.

PART-B: PROFILE OF NANDI DAIRY

Beginning of Nandi Group ofgggg Industries

Rayalaseema economically backward area in Andhra Pradesh was rarefied for industries. A dynamic entrepreneur Sri S.P.Y. Reddy who is basically mechanical engineer started his career with Nandi pipes. It is located in Nandyal at Kurnool district, a prominent town in Kurnool district. Initially the unit started producing black pipes way back 1977, strong determination and hard work Reddy come out successfully the finance assistance from the local commercial banks over

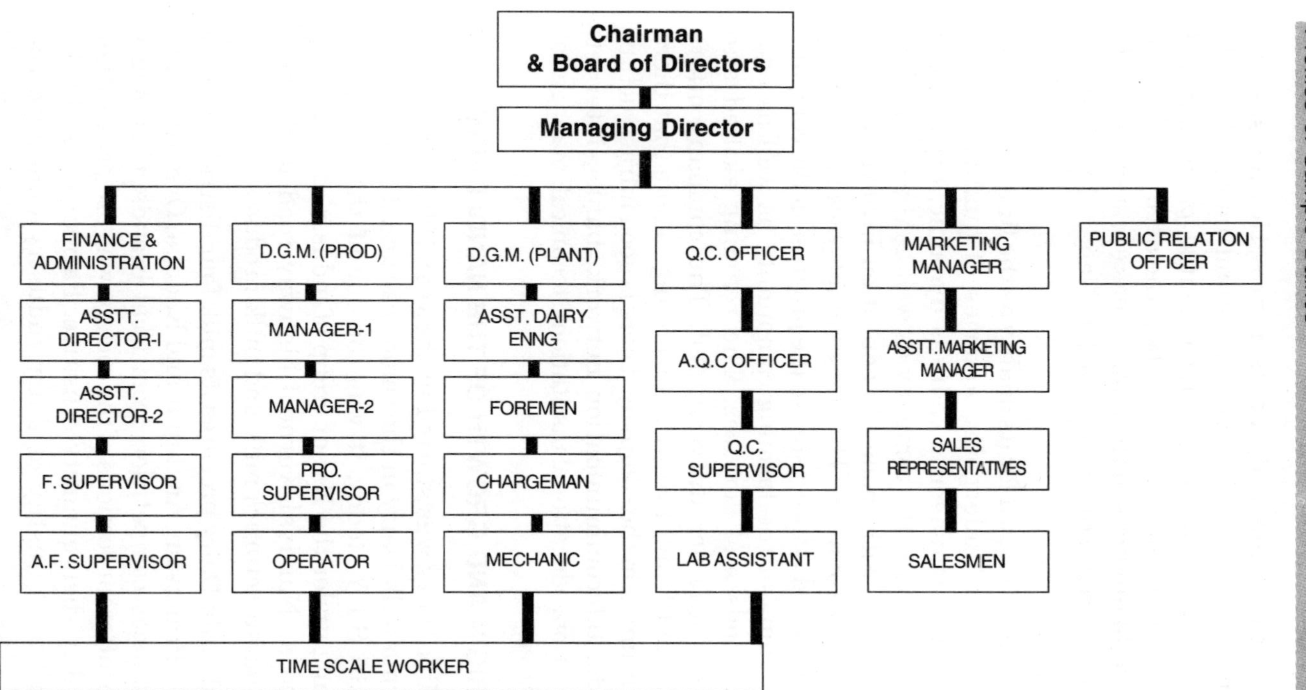

Source : Complied from the records of Vijaya dairy.

Chart 3.1. Organizational Chart of Vijaya Dairy

the company witnessed seculars growth and started marketing of pipes throughout its brand name is widely popular in all the locations and initiated procuring and distributing milk and milk products to the needy public. The milk and milk products in the name of piece erstwhile Nandi. Therefore, the company was named as Nandi Milk Products Pvt. Ltd.

The brand is known for its quality as the finished product turns out after undergoing a series of quality control technique. The managerial cadre is dynamic, experienced and has concern for the company as well as the public. The supervisor staff with moderate education involves actively in affairs of the company. This shows employees hailing from the local area and is interested in the progress of the company and well being of the farming community as well as end-users of goods and services. The reaming skilled and unskilled staffs are duty-minded. The company offers facilities like PF and ESI. It is surprising to note that there is no trade union in the company. There is good infrastructure like roads, telecommunication network, banks, internet, transportation, educational institutions, technical, managerial and electricity.

ORIGIN AND GROWTH OF THE NANDI DAIRY

Nandi Milk Dairy was started in the year 1997. It acquired from Nirma milk Dairy in the same year and it is under the care of Sri S.P.Y Reddy, it was the one of the industries functioning under the Nandi group of industries. This unit, is located at Nandyal-Kurnool highway road in Kurnool district, manufacturers milk and milk products.

Nandi Milk Dairy procures its milk four districts namely Kurnoool, Prakasam, Anatapur and Kadapa. Over the years this unit faces competition from local brands. Customers belong to different sections of the society, who are price rather than quality then quantity sensitive. Farmer house-holds, who are the mainstay of dairy industry dominate the scene.

Table 3.10. Select Details of Milk Procurement by Nandi Dairy as on 31-12-2007

Sl. No.	Parameters of milk procurement	Nos.
1	No. of MPCs/ MPACs	342
2	No. of Procurement routes	12
3	No. of villages covered	324
4	Milk Producers	16845
5	No. of milk chilling centres/ bulk units	5
6	Transport vehicles own/rented	25

Source: Compiled from the records of Nandi dairy, Nandyal.

Table 3.10 displays procurement particulars of Nandi Dairy. This dairy in an average 52 milk producers. The dairy has 342 MPCS/MPACs covering 16845 producers giving an average of 50 producers per MPCS/MPACs procures milk from 16845 producers spread over 324 villages. The company uses 25 vehicles owned/rented for procuring milk through 12 routes the transport system facilities the dealers to maintain optimum levels of inventory.

Milk Procurement by Nandi Dairy

The daily and annual milk procurement Nandi Dairy during 1998-2007 are furnished in Table 3.11. The daily average milk procurement stood at 4000 litres in 1998 as compared to 42100 litres against 2007. There has been gradual progress in the intervening years. Further 153.66 lakh litres of milk were procured in 2007 as compared to 14.60 lakh litres in 1998, recording an increase by 13 times. It is heartening note than annual percentage changes in daily milk procurement have been of positive magnitude. The highest magnitude of annual percentage change was highest at 50 in 1999 and honest at 14 in 2001. In the basis of preceding analysis it can be concluded that daily and annual procurement of milk have continued upward trend in the reference period.

Table 3.11. Daily and Annual Procurement of Milk in Nandi Dairy: From 1998-2007

Sl. No	Year	Daily Procurement (in liters)	Annual Milk Procurement (in lakh liters)
1	1998	4000 (....)	14.60
2	1999	6000 (50.00)	21.90
3	2000	7000 (16.67)	25.55
4	2001	8000 (14.28)	29.20
5	2002	10000 (25.00)	36.50
6	2003	13000 (30.00)	47.45
7	2004	17000 (30.77)	62.05
8	2005	25000 (47.05)	91.25
9	2006	33000 (32.00)	120.45
10	2007	42100 (27.57)	153.66

Note: Figures in parentheses are annual percentages increases ones the preceding years.

Source: Compiled from the records of Nandi Dairy

Procurement Prices, Processing and Sales in Nandi Dairy

Table: 3.12 presents per litre milk procurement prices in Nandi dairy, which correspond to fat content percentage. Higher the fat the greater the price per litre paid. For a fat content of five per cent the procurement price per litre of

Table 3.12. Milk Procurement Prices in Nandi Dairy During 2007

Fat%	Milkprice per/litre	Fat%	Milkprice per/litre	Fat%	Milkprice per/litre
5.0	12.00	6.7	16.08	8.4	20.16
5.1	12.24	6.8	16.32	8.5	20.40
5.2	12.48	6.9	16.56	8.6	20.64
5.3	12.72	7.0	16.80	8.7	20.88
5.4	12.96	7.1	17.04	8.8	21.12
5.5	13.20	7.2	17.28	8.9	21.36
5.6	13.44	7.3	17.52	9.0	21.60
5.7	13.68	7.4	17.76	9.1	21.84
5.8	13.92	7.5	18.00	9.2	22.08
5.9	14.16	7.6	18.24	9.3	22.32
6.0	14.40	7.7	18.48	9.4	22.56
6.1	14.64	7.8	18.72	9.5	22.80
6.2	14.88	7.9	18.96	9.6	23.04
6.3	15.12	8.0	19.20	9.7	23.28
6.4	15.36	8.1	19.44	9.8	23.52
6.5	15.60	8.2	19.66	9.9	23.76
6.6	15.84	8.3	19.92	10.0	24.00
	X			8.00	15.60

of FAT % age = 8%
of price per ltr of milk = Rs 15.60.
Source: Compiled from the records of Nandi dairy

milk is Rs. 12, for every 0.1 per cent increase in fat content, price increases by 24 paise per litre. The average fat content is eight per cent and the average price is Rs. 15.60 per litre. When the fat content gets doubled from 5 per cent to 10 per cent, the price per litre also gets doubled from Rs. 12 per litre to Rs. 24 per litre. From the above analysis conclude this dairy fixes procurement price per litre it can be based on the percentage of fat content in the milk which is the pricing methods followed by all dairies uniformly in A.P. state. Solid Non-fat (SNF) to down revision, depending in sliding down of as shown in table: 3.6. If this SNF is 8.8 per cent are more it pays the prices already fixed. For every 0.1 per cent decline in SNF, milk price declines by Rs. 06. In other words if the SNF 8.7 per cent the amount deducted from the normal price per litre is 0.06 paisa. If Snf falls to the lowest level of 8 per cent, the maximum 48 paise will be deducted from the normal price per litre.

Table: 3.13. Price Reduction per Liters *vis-à-vis* Sliding Down SNF Per Cent

Sl. No.	Fat %	Milk price (Rs. paise)
1	8.8	Nil
2	8.7	0.06
3	8.6	0.12
4	8.5	0.18
5	8.4	0.24
6	8.3	0.30
7	8.2	0.36
8	8.1	0.42
9	8.0	0.48

Source: Compiled from the records of Nandi Dairy

In Table 3.13 are shown price per litre sliding down as the SNF percentage declines. To be specific, 0.1 per cent decline in SNF results in 6 paise reduction in price per litre. The SNF percentage sliding down to the maximum level of 8.8 to 8.0 attracts price reduction from 'o' paise to 48 paise per litre.

The milk collected is sent into bulk containers and chilled. After this milk is pumped into bulk tanks which contain an insulated body. Through this milk is sent to factory for processing, the processed milk is delivered to consumers. In the pasteurization process milk is heated at 80°C and later cooled at 4°C. This process removes 90 per cent of bacteria and the life of the milk increases to the minimum of two days. Homogenization process is carried out after pasteurization process; this process is under taken to bring uniform quality. It enhances taste and thickness of the curd. The processed milk is packed in polythin covers. Before packing quality and quantity are examined in the laboratory. The filled in polythin covers are kept in trays under cold storage till loaded for transportation.

As can be observed from Table 3.14 the dairy produces different types of by-products such as *Khoya*, kalakan, *paneer*, *doodhpeda*, milk *mysore* pack, *makhan*, *ghee* and butter milk informally in all the sweet products fat content is 26 per cent and SNF content 38 per cent with a shelf life of five days. In the case of ghee by-products fat content is 99 per cent whereas in buttermilk it is 1.4 per cent the percentage of SNF in butter

Table 3.14. Price Rates of By-products of Nandi Dairy

Sl.No.	Name of the product	Quantity (in kg/ml)	Product price (Rs. per/kg)	Fat %	SNF%	Shelf life
1	*Khoya*	Kg	120	26	38	5 days
2	Kalakan	Kg	120	26	38	"
3	Paneer	Kg	120	26	38	"
4	Doodh peda	Kg	120	26	38	"
5	Milk *mysur pack*	Kg	120	26	38	"
6	*Makhan*	Kg	120	26	38	"
7	*Ghee* (Per/kg)	Kg	160	99	..	6 months
8	Butter Milk (200ml pockets)	200ml	4	1.4	4	One day

Source: Compiled from the records of Nandi Dairy.

milk is four per cent. The shelf life for ghee is six months and butter milk one day. All by products are highly percentages since they short shelf life language from 1 day to 5 days, the notable exception being *ghee* which has a shelf life of six months.

The daily and annual sales in Nandi Dairy in the reference period, 1998-2007 are furnished in Table 3.15. The daily sales were 15000 litres in the initial year which

Table 3.15. Daily and Annual Milk Sales of Nandi Dairy: From 1998-2007

Sl. No	Year	Daily sales (litres)	Annual sales (in lakh litres)
1	1998	15000 (...)	54.75
2	1999	14896 (-0.09)	54.70
3	2000	20000 (33.45)	73.00
4	2001	25000 (25.00)	91.25
5	2002	26000 (4.00)	94.90
6	2003	30000 (15.38)	109.50
7	2004	32000 (6.67)	116.80
8	2005	36000 (12.50)	131.40
9	2006	38000 (5.55)	138.70
10	2007	43000 (13.15)	156.95

Note: Figures in parentheses are annual percentages increases ones the preceding years.

Source: Compiled from the records of Nandi Dairy

gradually increased to 43000 litres in 2007, terminal years, and registering 2.86 times growth. A similar trend is observed with regard to annual sales were 54.75 lakh litres in 1998 and 156.95 lakh litres in 2007. Figures in parentheses in column three of the table highlight the annual percentage increases over the preceding years. Except in which negative growth rate is recorded in 1999, the annual percentage increases are positive for the rest of the period. This positive figure is lowest at four per cent in 2002 and highest at 33.45 per cent in 2000.

Channels of Distribution of Nandi Dairy

The Nandi Dairy physically distributes the milk and milk products through the different channels. The company employees direct marketing without any intermediaries in the channel of distribution between company and consumer. It distributes its products to educational institutions, hospitals, hotels, oil mills, etc. directly without involving any intermediary in between them. Another channel employed is indirect marketing in which the dairy distributes it's to products to dealers and dealers in turn sell them to customers. There are 310 dealers of the company in the state of AP. Further, the company has established its own retail outlets in Rayalaseema region of AP. Through which products are supplied to the customers.

Sales of Milk Products, By-products and Products Prices of Nandi Dairy

As can be seen from Table 3.7 that all types of milk are sold in three packages like that litre(1000ml), 500ml and 200ml. Price depends upon the fat and snf content. The skimmed milk is sold for Rs. 15 per litre, Rs.7 per 500ml and Rs. 3 per 200 ml. Toned milk is sold for Rs.18, Rs. 9.50 and Rs. 4 respectively for the aforesaid quantities. Doubled tone milk is priced at Rs.17, Rs.8.50 and Rs. 3 per litre and half liter and 200ml, Toned milk, double toned milk, standardised milk like this are also packed with three different quantities to

Table 3.16. Price Rates of Different Types of Milk Products in Nandi Dairy

Sl. No	Milk Products	Price	Fat %	Snf %	Shelf life
1	Skimmed Milk	1000ml : Rs.15 500ml : Rs.7.50 200ml : Rs.3.00	0.1	8.7	One day at shelf life
2	Double Toned Milk	1000ml : Rs.17.00 500ml : Rs.8.50 200ml : Rs.3.50	1.5	9	One day at shelf life
3	Toned Milk	1000ml : Rs.18.00 500ml : Rs.9.50 200ml : Rs.4.00	3	8.5	One day at shelf life
4	Standerdised Milk	1000ml : Rs.20.00 500ml : Rs.10.00 200ml : Rs.4.00	4.5	8.7	One day at shelf life
5	Whole Milk	This milk is sold as: 1000ml : Rs.22.00 500ml : Rs.11.00	6	9	One day at shelf life

Source: Primary Data

down-reach the market. The skimmed milk is sold at Rs. 17, 7, 3 for packages of different quantities in that order respectively. The toned milk contains three per cent and 8.5 per cent fat and SNF contents sequentially. The milk is sold at Rs. 18, Rs. 9.50, and 4 per packet with a quantity of 1000ml, 500ml and 200ml. In case of standardised milk is quantities are sold at price of Rs. 20, Rs. 10 and Rs.4 sequentially. The whole/full cream milk is sold at Rs.22 per litre and Rs. 11 per half litre. Each of the sweet products is sold Rs.120 per kg. In the case of by products, ghee per kg fixed Rs.160 and 200ml buttermilk at Rs.4. Based on the fat and SNF contents, the company fixes price for its products. Reading column four of the table it can be noticed that the higher the fat content in the different milk products, the higher the prices charged. The entire milk products have a common shelf life of one day.

The Nandi Dairy sells sweet products such as *khoya*, *kalakhan*, *doodhpeda*, milk *mysur pak*, *palakova* and *makhan* for the reference period 2000-2007, sales figured are furnished in Table 3.17. The sale of khoya as gradually increased from 15175 kg in 2000 to 54666 kg in 2007 except a marginal decline in 2006. The increase of sales in 2007 over 2006 sales is remarkable. A similar trend can be formed in the sales of *kalakhan, doodhpeda*, milk *mysur pak, palakova*, *makhan* and *doodhpeda*. Reading column three of the take it can be noticed that percentage shares all milk products in total sales are of almost equal magnitude. To be specifying *khoya*, *kalahgan*, *doodhpeda, milkmysurpak*, *plakova* and *makhan* had percentage shares of 17.10, 15.16, 15.45, 18.41, 17.33 and 16.53 respectively in that order in 2000. The structure of percentage shares of those products as obtained in 2007, when compared to 2000, remains unchanged.

Sale figures of by-products Nandi Dairy for the period 1998 to 2007 are furnished in Table 3.18. It can be seen from table that there is gradual increase in the quantum of sale of *ghee* from 6000 kg in 1998 to 13000 kg in 2007, regarding 116.67 per cent increase. With regard to buttermilk 2.34 lakh packets were sold in 1998 and with progressively increased 8.55 lakh packets in 2007, registering 265.78 per cent increase. In the objective evidence of apart in sales of the two by products, it can be concluded that Nandi dairy recorded secular up-trend in the sales of the by products in the reference period.

Organization Chart of Nandi Dairy

Top-down organization chart of Nandi dairy is presented in Chart 3.2. Board of Directors headed by chairman, is plural top executive involved in policy-making. Executive Director and functional directors who look in the next layers after finance, purchases, production marketing and public relations are shown. There are managers shown in descending order. At present most of the funds are obtained, through self financing except a loan on hypothecation of stock from SBI Nandyal branch. Initially external sources were accused. Finance department headed by financial

Table 3.17. Sweet Product Sales of Nandi Dairy : From 2000 to 2007

(in kgs.)

Sl.No.	Particulars	Years							
		2000	2001	2002	2003	2004	2005	2006	2007
1.	Khoya	15175 (17.10)	23783 (16.06)	29541 (16.58)	30558 (16.67)	34250 (16.46)	38208 (16.35)	36508 (16.03)	54666 (15.29)
2.	Kalakhan	13450 (15.16)	26208 (17.70)	31541 (17.70)	29150 (15.90)	33916 (16.30)	36100 (15.45)	34053 (14.96)	64816 (18.13)
3.	Doodhpeda	13708 (15.45)	24700 (16.68)	30500 (17.11)	31950 (17.43)	32950 (15.84)	37566 (16.08)	37450 (16.45)	54958 (15.37)
4.	Milk *mysurpak*	16333 (18.41)	25125 (16.97)	28625 (16.06)	30183 (16.47)	35483 (17.06)	39416 (16.87)	39108 (17.18)	63675 (17.81)
5.	*Palakova*	15375 (17.33)	25916 (17.50)	29500 (16.55)	29983 (16.36)	34550 (16.61)	39983 (17.11)	40200 (17.66)	54800 (15.33)
6.	*Makhan*	14666 (16.53)	23000 (15.53)	28458 (15.97)	31408 (17.41)	36813 (17.70)	42308 (18.11)	41208 (18.10)	64533 (18.05)
	Total	88707 (100.00)	148012 (100.00)	178165 (100.00)	183232 (100.00)	207962 (100.00)	233581 (100.00)	227627 (100.00)	357448 (100.00)

Note: Figures in parantheses are percentage sto respective column totals.
Source: Primary Data.

Table 3.18. Sales of By-products in Nandi Dairy during the Year 1998-2007

(in kgs.)

Sl.No.	Particulars	Years									
		1998	1999	2000	2001	2002	2003	2004	2005	2006	2007
1.	Ghee (Kg)	6000	6800	7100	7400	7500	8100	8300	9600	9625	13000
2.	Butter-milk (Pockets)	2,33750	2,38750	3,05000	4,37500	4,45250	4,87500	4,66000	5,01500	5,15000	8,55000

Note: Figures in parantheses are percentage sto respective column totals.
Source: Primary Data.

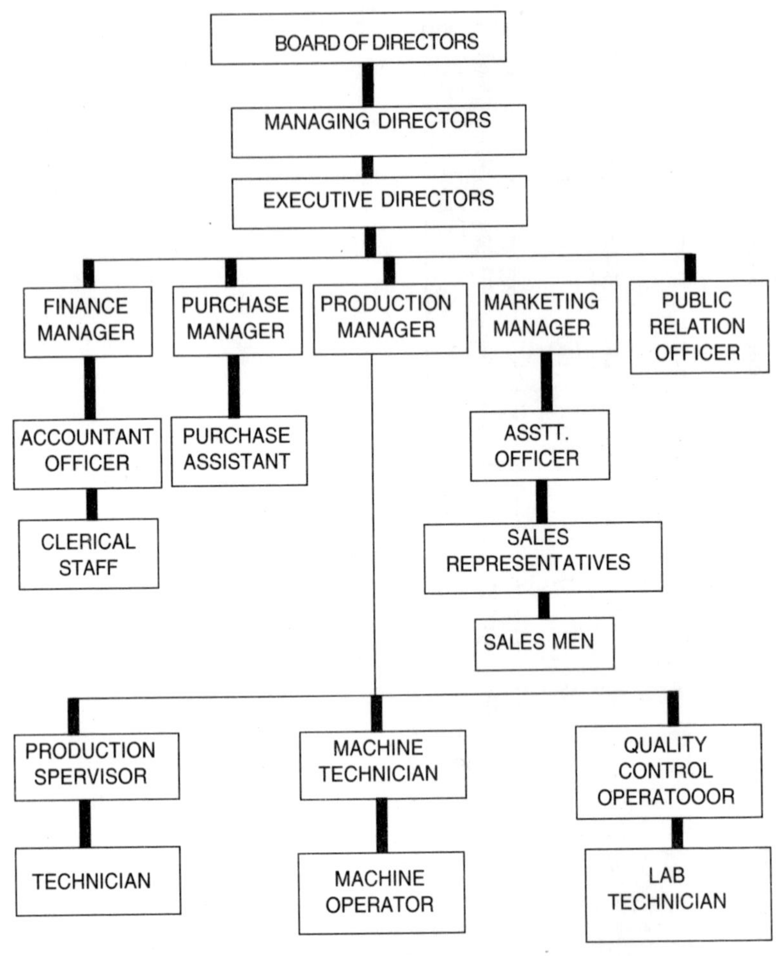

Chart 3.2. Organizational Charts of Nandi Dairy

manager who is in-turn is assisted by four accountants and four junior assistants. The company follows a policy of cash and carry system in selling its output. Marketing department is under the control of marketing manager assisted by assistant marketing manager, 20 sales men and 30 sales representatives. The company affects sales by formulating and implementing marketing mix. The human resources are looked after by HR manager, the primary task of being recruitment, selection, placement and induction compensation safety, health care and boarding scenes during working hours. In addition there are public relations and quality control departments.

Conclusion

KDMPMACU is cooperative dairy operating in Kurnool district. It procures milk from 36 thousand milk producers spread over 566 villages in Kurnool and Cuddapah districts annual procurement of milk 2007 stood at 322.18 lakh litres. Milk procurement prices are fixed based on fat and SNF content. All milk products have common shelf life of one day, making them highly perishable products. Annual sales of milk stood at 192.95 litres in 2007. There has been chasing trend in sales of byproducts except ghee.

Nandi diary, private sector diary was established in 1997. It procures milk 16845 milk producers spread over 324 villages. Annul procurement milk in Nandi diary was 153.66 laksh liters in 2007. Milk by products have shelf life of 5 days except buttermilk having one day shelf life and ghee having six month shelf life. Sweet products sold by Nandi dairy are of equal importance of these percentages in total sales in kg units. Two important by products sold by dairy are ghee and butter milk.

4 CHAPTER

TRENDS IN MILK PROCUREMENT IN SAMPLE DAIRIES

INTRODUCTION

In this chapter, an attempt is made to examine various issues relating to procurement of milk in terms of quantities and values daily, seasonally. There are focused on the two sample dairies, Vijaya and Nandi. In addition, capacity utilization in milk chilling centers of these two dairies is studied as well.

Milk Procurement and Production in Vijaya Dairy

As referred earlier, Vijaya Dairy is a cooperative unit with substantial presence in milk procurement and production. Table 4.1 milk procurement in the rural areas for individual years. Milk production takes place in small quantities. After meeting their daily needs, the marketable surplus milk is sold to village Milk Cooperative Societies/Milk Collection Agents (MCS/MCAs) twice a day i.e., morning and evening. MCS/MCAs after meeting local demand, supply remaining milk are supplied to Vijaya dairy. The details of MCS/MCAs under Vijaya dairy for the period 1998-2007 are provided in Table 4.1. The total procurement of milk for the year 1998 from all societies was 92.08 lakh liters as against 322.18 lakh liters in 2007. There are ups and downs in the yearly increment. For example the increase was in the range of 8.10 to 58.55 per cent, whereas it declined by 23.61 per cent in 2003 and 13.75 per cent in 2006. There is a decline by 23.60 per cent and 13.74 per cent during in 2003 and 2006 respectively. The daily average milk procurement for society worked out to be 52.22 liters in 1998 while it was 122.76 liters in 2007. In the mean time, there are wide variations in it. For example, 2005 registered the highest at 133.33 liters and 1999 the longest at 48.35 litres. The reason for volatile trend is due to fluctuations in the number of MCS/ MCAs and daily procurement of milk. The reasons for these are not far to seek it may be concluded that there is a growth in the number of MCS/MCAs, daily average milk procurement and procurement per society during the period under review with marginal fluctuations.

Table 4.1. Milk Procurement in Vijaya Dairy: From 1998 to 2007

Year	No. of societies/ milk collection agents	Daily average milk procurement (liters)	% ge of annual change	Daily average milk procurement per society	Annual milk production (lakh litres)
1998	480	25227.39	.....	52.55	92.08
1999	625	30224.65	19.80	48.35	110.32
2000	680	44706.84	47.91	65.74	163.18
2001	620	56339.72	26.01	90.87	205.64
2002	715	89331.50	58.55	124.93	326.06
2003	645	68241.09	-23.60	105.80	249.08
2004	711	79986.30	17.21	112.49	291.95
2005	710	94665.75	18.35	133.33	345.53
2006	735	81649.31	-13.74	111.08	298.02
2007	719	88268.49	8.10	122.76	322.18
	X	65864.15		96.79	240.40
	SD	25377.19		30.97	92.63
	CV	38.52		31.99	38.53

Source: Primary data.

Trends in Milk Procurement and Production in Nandi Dairy

As referred earlier Nandi Dairy in major private player in the dairy industry in Kurnool District which has been recording up-tend in procurement and production of milk. The particulars of milk procurement in Nandi Dairy during 1998-2007 is furnished in Table 4.2. In Nandi Dairy, 139 societies supplied milk to it in 1998, which increased to 342 MCS/MCAs in 2007. The daily average milk procurement stood at 4040 litres

Table: 4.2. Milk Procurement in Nandi Dairy during 1988-2007

Year	No. of societies/ milk collection agents	Daily average milk procurement (litres)	% age of annual change	Daily average milk procurement per society	Annual milk production (lakh litres)
1998	139	4040	...	29.06	14.60
1999	130	6032	49.30	46.40	21.90
2000	120	7152	18.57	59.60	25.55
2001	140	8650	20.95	61.78	29.20
2002	153	10025	15.90	65.52	36.50
2003	145	13320	32.87	91.86	47.45
2004	180	17615	32.24	97.86	62.05
2005	250	25034	42.11	100.13	91.25
2006	294	33065	32.09	112.46	120.45
2007	342	35655	7.83	104.25	153.66
	X	16058.80		76.89	60.26
	SD	11447.05		28.07	46.86
	CV	71.28		36.50	77.82

Source: Primary Data.

in 1998 as compared to 35655 litres in 2007. There is a progressive growth in the milk procurement during the period said. In other words there is no decline in any of the years. The rate of yearly growth, however, varied between the longest 7.83 per cent in 2007 to highest 47.30 per cent in 1999. Mean while there are ups and downs in milk procurement which an average per day per society was 104.45 liters in 2007 as against 29.06 litres in 1998. It may be noted that there is a highest procurement was at 122.46 litres in 2006. It can be summed up that the number of MCS/ MCAs that supplied milk to Nandi Dairy fluctuated widely despite a persistent increase. Through there is an upward trend in the daily average milk procurement; there are significant variations, in the annual increments. The yearly procurement was 14.60 lakh litres in 1998 as compared to 153.66 lakh liters during 2007. There is a gradual and consistent increase in the yearly procurement in the reference period. However, there are fluctuations in the annual increment. For example the year 2007 reported the lowest growth at 7.83 per cent whereas the 1999 the highest at 49.30 per cent. There are noticeable variations in the procurement of milk between Vijaya and Nandi Dairies, the reason being fluctuations in the number MCS/MCAs, daily average milk procurement, annual growth daily average milk procurement per society. Procurement in Nandi dairy may be favorable compared with Vijaya Dairy, because there has been a continuous upward trend in the latter. The Vijaya Dairy is under co-operative sector and managed by officials, whereas Nandi Dairy is in the private sector and therefore, there is a personal element in the management of its affairs. It may be concluded that the performance of Nandi Dairy is better than in the Vijaya Dairy.

Lean and Peak Seasonal Variations in Milk Procurement in Vijaya Dairy

Seasonal variations in milk procurement are conspicuous in dairy industry, which is captured in what fallows.

Table 4.3. Daily Average Milk Procurement in Flush and Lean Season in Vijaya Dairy: From 1998-2007

Sl. No.	Dairy average procurement		col (3) as %ge of col (2)
	Flush season (in liters)	Lean season (in liters)	
1998	14884.00	10343.00	69
1999	17530.00	12694.65	72
2000	25035.83	19671.01	79
2001	30986.84	25352.88	82
2002	54492.21	34839.29	64
2003	38897.42	29343.67	75
2004	46392.05	33594.25	72
2005	58692.16	35973.59	61
2006	47356.70	34292.61	72
2007	54726.16	33542.33	61
X	38899.34	26964.73	
SD	15990.41	9576.90	
CV	41.10	35.51	

Source: Primary Data.

Table 4.3 exhibits the dairy milk procurement by Vijaya Dairy during flush, lean seasons. Dairy milk procurement in flush season was 14804 litres in 1998, which gradually up trended in subsequent years to reach second highest level of 54492 litres in 2002. Since then daily procurement down trend in subsequent years to reach all-time high of 54726 litres in 2007. Daily milk procurement in lean season was 10343 litres in 1998 which started climbing up to reach second high level of 34839 litres in 2002. Since then it started declining in the next two years to reach all time peak of 35973 litres in 2005. In the next two years there was a down ward trend. Focusing attention on column four, it can be witnessed that the daily average milk procurement in lean

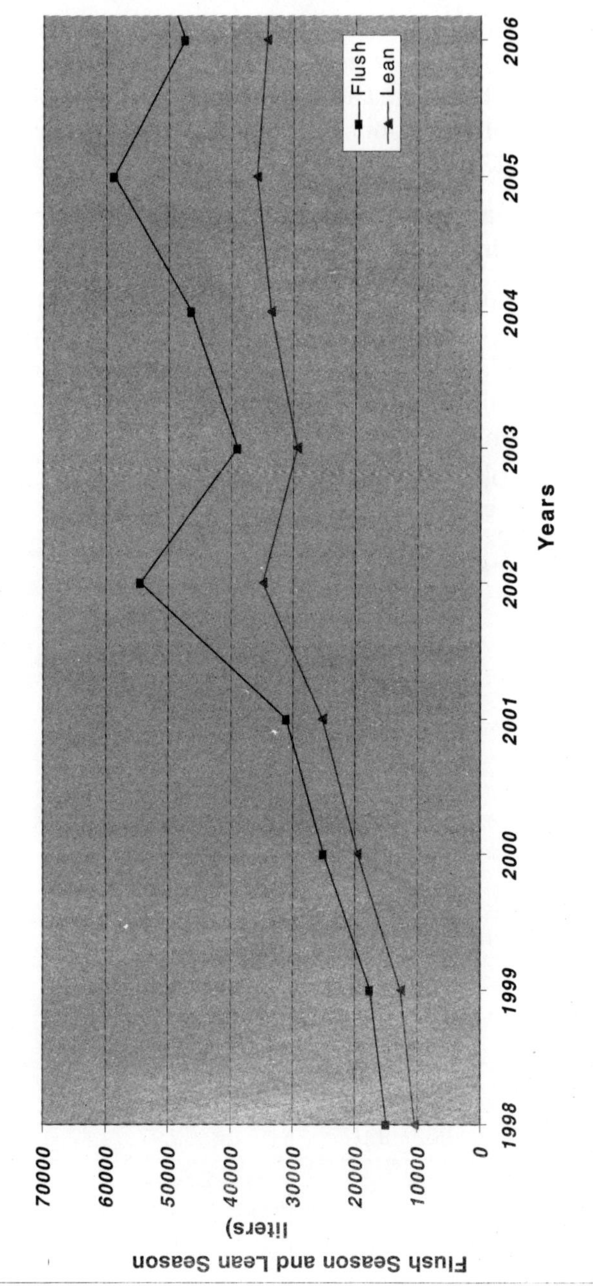

Fig. 4.1. Daily Average Milk Procurement in Flush and Lean Seasons in Vijaya Dairy: From 1998-2007

season as percentage of flush season was longest at 61 per cent in 2005 and 2007 as compared to the highest of 82 per cent in 2001. Based on the above analysis, it can be concluded that seasonal variations are marked in daily average milk procurement.

Lean and Peak Seasonal Fluctuations in Milk Procurement in Nandi Dairy

Table 4.4 displays daily average milk procurement by Nandi Dairy during flush and lean seasons for 10 year period from 1998 to 2007. In 1998 the daily average milk procurement in flush season was 2302 litres which kept increasing years after year since

Table 4.4. Daily Average Milk Procurement in Flush and Lean Seasonal in Nandi Dairy: From1998-2007

Sl. No.	Dairy average procurement		col (3) as %ge of col (2)
	Flush season (in liters)	Lean season (in liters)	
1998	2302.00	1898.00	82
1999	3679.00	2353.00	64
2000	4219.68	2932.32	69
2001	5363.00	3287.00	61
2002	5914.00	4111.00	70
2003	8125.20	5194.80	64
2004	10216.70	7398.30	72
2005	15270.74	9763.26	64
2006	19508.35	13556.65	69
2007	21749.55	13905.45	64
X	9634.82	6439.97	
SD	*6906.46*	*4532.54*	
CV	*71.68*	*70.38*	

Source: Primary Data.

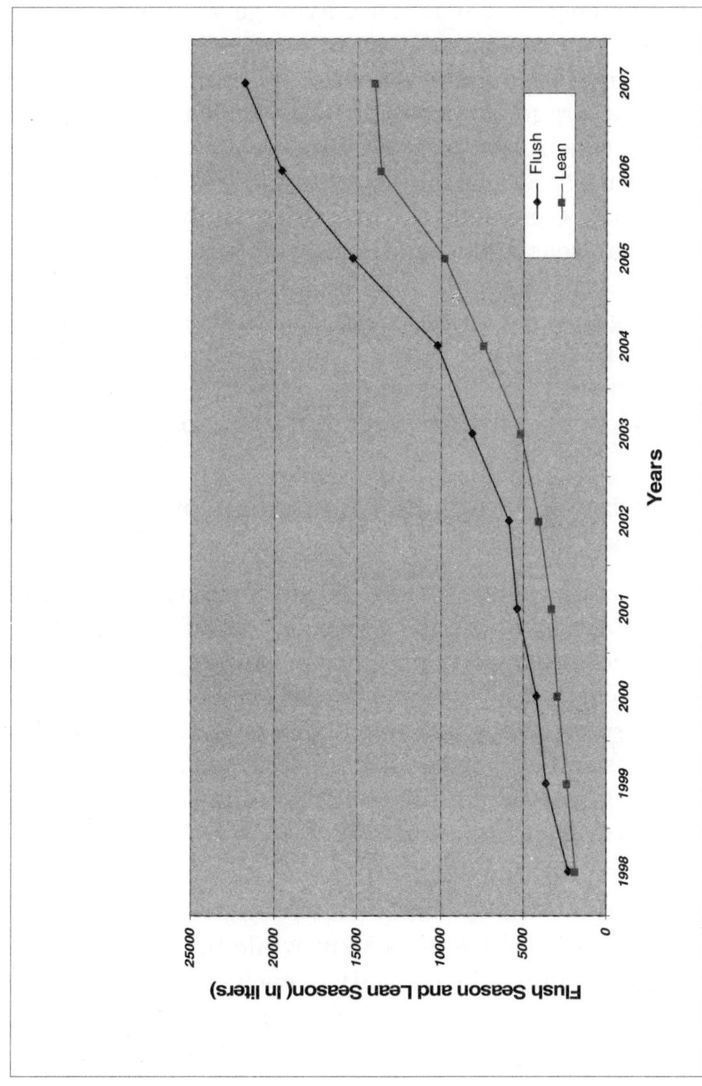

Fig. 4.2. Daily Average Milk Procurement in Flush and Lean Season in Nandi Dairy: From 1998-2007

Fig. 4.4 then to reach the peak level of 21750 liters in 2007. Similar up-trending of daily procurement of milk during lean season can be observed. The daily average was 1898 liters which climbed up consistently by to reach all time peak of 13905 liters in 2007. Paying attention to column for which shows daily average procurement during lean season as percentage of that during flush season, it can be noticed that it was peak with at 82 per cent in 1998 and the longest at 61 per cent in 2001. Guided by the preceding analysis it can be concluded that seasonal variations are conscious in daily average milk procurement. Joint reading of the table and the preceding one it can be concluded that average milk procurement in flush and lean seasons in Nandi dairy up trended consistently in the reference period, as compared to that of Vijaya dairy which experienced ups and downs year after year.

Capacity Utilization in Milk Chilling Centers of Vijaya and Nandi Dairies

Capacity utilization impacts cost of production, higher the capacity utilization lower the average cost of production. Hence, the focus on capacity utilization. Chilling centre-wise installed capacity, utilized capacity and capacity utilization of Vijaya Dairy all shown in Table 4.5 for the year 2007. The dairy procures milk from 16 chilling centres. The distance between the dairy and chilling centres varies considerably longest 210 km from Rajampeta and shortest 53 km from Dhone.

For example the distance between the dairy and Rajampeta chilling centre is 210 km while 53 km between Dhone and dairy unit. The capacity of each of three chilling centres is 4000 litres, such as Kolimigundla Atmakur, Dornipadu and Dhone 4000 litres and Pattikonda 4500 litres. The capacity each of 9 chilling centres such as Banavasi, Halaharvi, Sankalapuram, Chagalamarri, Mydukur, Rayachoti, Pulivendula, Proddutur, Dornipadu and Rajampeta is 5000 litres per day, the Kurnool is are

Table: 4.5. Daily Average Capacity Utilization in Milk Chilling Centres of Vijaya Dairy During 2007

Sl. No	Centre name of the chilling cente	Dairy centre distance (in km.)	Installed capacity (in litres)	Average daily utilization capacity (in litres)	Col 5 as % ge of col 4
1	Banavasi	90	5000	3500	70.00
2	Halaharvi	95	5000	3200	64.00
3	Patti konda	74	4500	2850	63.33
4	Sankalapuram	85	5000	3800	76.00
5	Kolimigundla	90	4000	3150	78.75
6	Chagalamarri	60	5000	4500	90.00
7	Mydukur	75	5000	3900	78.00
8	Rayachoti	170	5000	2850	57.00
9	Pulivendula	180	5000	3100	62.00
10	Proddutur	85	5000	4200	84.00
11	Atmakur	75	4000	3600	90.00
12	Dornipadu	60	4000	3400	85.00
13	Dhone	53	4000	3550	88.75
14	Rajampeta	210	5000	4800	96.00
15	Kurnool	78	50000	16550	33.10
16	Nandyal	....	100000	19318	19.61
	Total		216500	88268	40.77

Source: Compiled from the records of Vijaya Dairy, Nandyal.

moderately milk (20000 litres) while Nandyal very big one lakh liters against the installed capacity the chilling units procured milk has follows two units such as Pattikonda and Rayachoti each procured 2850 liters, on an average per day. The Kurnool and Nandyal chilling units procured 6550 liters and 19381 liters respectively. There are nine chilling units in the range of 3000-4000 litres. The total installed capacity of all the chilling centers is put together 216500 liters per day. The milk actually collected was 88268 liters registering, a capacity utilization of 40.77 per cent at the aggregate level the capacity utilization across the chilling centers, differ remarkably. It is sad to know that the capacity utilization is very low at 19 per cent Nandyal chilling centre while the highest in the Rajampeta chilling centre at 96 per cent. The level of accomplishment is 80-90 in six centers, 70-80 per cent in three centers 60-70 in four centers and 33 per cent in Kurnool centre. It may be concluded that Vijaya dairy collects milk from distance bases spread over Kurnool and Kadapa districts. Further the capacity utilization is poor, further more it is very poor in Nandyal and Kurnool centers. The capacity utilization is low due to the nascent growth of private dairies such as, Heritage, Balaji, Kamadhenu and Nandi in around the Nandyal.

Table 4.6. Daily Average Capacity Utilization in Milk Chilling Centers of Nandi Dairy During-2007

Sl. No	Centre name of the chilling cente	Dairy centre distance (in km.)	Installed capacity (in litres)	Average daily utilization capacity (in litres)	Col 5 as % ge of col 4
1	Mydukur	75	10000	8000	80.00
2	Jammulamadugu	65	8500	7000	82.35
3	Bestavaripeta	71	6000	5000	83.33
4	Yadiki	80	7500	5500	73.33
5	Nandikotkur	65	3000	2500	83.33
6	Nandyal	....	15000	7655	51.03
	Total		50000	35655	71.31

Source: Compiled from the records of Nandi Dairy, Nandyal.

Table 4.6 Modeed on the proceeding table provides data on daily milk chilling capacity utilization in the chilling centers of Nandi dairy unit for the period of 2007. Thus, dairy has six chilling centres with varying installed capacity the lowest being 3000 liters in Nandikotkur and the highest being 15000 liters in Nandyal. The aggregate installed capacity for all centres put together was 50 thousand litres and the aggregate capacity utilized was 35655 liters. Paying attention to column 6 of the table, it can be noticed that overall capacity utilization of the dairy was 71.31 per cent. Compared to the overall average, all the centers expecting Nandyal center had higher utilization rates. The highest utilization recorded was 83.33 per cent in Bestavaripeta and Nandikotkur dairies and the lowest being 51.03 per cent in Nandayal. Combained reading of this and the preceding tables, it can be concluded that in daily average milk chilling capacity utilization Nandi dairy has a better record than Vijaya dairy unit.

Conclusion

Milk procurement and production up trended in the reference period in both Vijaya and Nandi dairies. Value of milk procured lean season is two-thirds of procured volume in flush season, which is the marked feature of dairy industry. In terms of capacity utilization wide variations are noticed among chilling centers of both the dairies.

5 CHAPTER

PRODUCT RELATED MARKETING PRACTICES

Introduction

This chapter portrays marketing practices related to product aspects of Vijaya and Nandi dairies, following the 4 P's conceptual framework of marketing. Accordingly this chapter throws light on dairy products hierarchy, product lines product depth, packaging, ingredient branding, product line stretching and distribution of sales by product lines.

Product of Sample Dairy Units

Three product lines of each dairy are discussed here under Table 5.1 list out individual products of Vijaya and Nandi dairy units under different product lines concentrating on column two of the table observed Vijaya dairy has three product lines namely milk product, basic product and by-products of the three product lines, by product line is the longest product line with six individual product of followed by milk product line with five products and basic with single product. In terms of total number of individual products, it has twelve products that is its depth of product line. Observing column three of the table, it can be found that Nandi Dairy also has three product lines, sweet product line is the longest six products, followed milk product line with five products and by products with product in terms of product depth, that is, total number of individual product in three different product lines is 13 products. In comparison both the dairy units are same in terms of product width then coming to product depth in Nandi Dairy has a marginal superiority over Vijaya Dairy unit. Thus, both the dairies appear to have a similar scope in terms of product line product width and product depth.

Table: 5.1. Products of Vijaya Dairy and Nandi Dairy

Sl. No.	Products of Vijaya Dairy	Products of Nandi Dairy
1.	(1) Milk Product Line (a) Skimmed Milk (b) Double Toned Milk (c) Toned Milk (d) Standardized Milk (e) Gold Milk	(1) Milk Product Line (a) Skimmed Milk (b) Double Toned Milk (c) Toned Milk (d) Standardized Milk (e) Whole Milk
	(2) Basic Product Line (a) Paneer	(2) By-product Line (a) Ghee (b) Butter-milk
	(3) By Product Line (a) Butter (b) Butter Milk (c) Ghee (d) Skimmed milk powder (e) Doodh peda (f) Sterilized flavoured milk	(3) Sweet Product Line (a) Khoya (b) Kalakan (c) Doodh peda (d) Milk Mysur Pack (e) Palakova (f) Makhan

Note: Individual products are categorised under different product lines.
Source: Primary Data.

Product Hierarchy

Product hierarchy consists of five-layer classification of dairy products. Table 5.2 presents product hierarchy in sample dairy units.

This hierarchy ranges from basic needs to particular items which satisfy those needs. Accordingly six levels of the product hierarchy are identified the dairies. Individually each of these levels is explained below.

The first level is need family. In this level dairy product satisfy core need that underlies the existence of dairy products; on example food products. The second level is

Table 5.2. Dairy Product Hierarchy in Sample Dairies

Sl. No.	Dairy product hierarchy	Vijaya Dairy	Nandi Dairy
1.	Need Family	Food Products	Food Product
2.	Product Family	Dairy Products	Dairy Products
3.	Product Class/ Categories	Processed Dairy Products	Processed Dairy Products
4.	Product line	(1) Milk Products (2) Basic Product (3) By Products	(1) Milk Products (2) By Products (3) Sweet Product
5.	Product type/Product items	(1) Milk Products (a) Skimmed Milk (b) Double Toned Milk (c) Toned Milk (d) Standardized Milk (e) Gold Milk	(1) Milk Products (a) Skimmed Milk (b) Double Toned Milk (c) Toned Milk (d) Standardizd Milk (e) Whole Milk
		(2) Basic Product (a) Paneer	(2) By products (a) Ghee (b) Butter-milk
		(3) By Products (a) Butter (b) Butter Milk (c) Ghee Product (d) Skimmmed milk powder (e) Doodh peda (f) Sterilized flavoured milk	(3) Sweet Product (a) Khoya (b) Kalakan (c) Doodh peda (d) Milk Mysur Pack (e) Palakova (f) Makhan

Source: Primary data.

product family. All dairy products that satisfy core need with effectiveness belong to product family. Third level is product class or category. Group of products within the product family recognized as having certain functional coherence belong to product class. Processed dairy products belong to product class or category. Forth level is product line in the sample dairy units. It is a group of products that are closely related perform similar function and are sold to the same customer group, marketed through the same outlets or channels and fall within the given price ranges. There are three product lines in the dairy units on Vijaya Co-operative Dairy the three product lines are milk products, basic products and by-products. In Nandi Dairy also three product lines are noticed such as milk product, by-product and sweet products. The next level in the hierarchy is product type a group of items within a product line that share one of the several possibly forms of the product. The sixth level is items - a distinct unit within a brand or product. The fifth and sixth levels in the sample dairy units are as following on Vijaya Co-operative Dairy skimmed milk, toned milk double toned milk, standardized milk and whole milk belong to milk type individual products. Pannier is another individual product. Buttermilk, *ghee* product, skimmed milk powder, *doodhpeda*, sterilized flavoured milk are individual products in by-product lines. In Nandi Dairy skimmed milk, toned milk, double toned milk, standardised milk and gold milk are individual products in milk product line. *Ghee* and butter milk are only two individual products in by-product line. As compared to Vijaya Co-operative Dairy and Nandi dairy has additionally sweet product line in *khoya, kalkhan, doodhpeda, milk mysur pack, palkova* and *makhan* are individual products. From the processing analysis it can be noticed that dairies have six levels each in product line each. A prominent product line in Nandi dairy is sweet products having six individual products, which is not there is Vijaya Co-operative Dairy. In basic product line pannier is the individual product marketed by Vijaya dairy only.

Product-line Sales in Sample Dairy

Sales with reference to 2007 year are discussed along with the product lines for both the dairies. In Table 5.3 product-mix width and length of Vijaya dairy the sales of each product line and individual products under each product line are shown in absolute values and percentage terms for the year 2006-07. The absolute sales value were Rs. 3,858.10 lakhs of milk product line, Rs.10 lakh of basic product line and Rs. 176.41 lakh of by-product line. Unit aggregate sales were value Rs. 4044.68 lakh, in the last column of the table are shown percentage share of each individual product in total sales volume. Milk products accounted for 95.39 per cent of total sales value, followed by by-products line with percentage share of 4.36 per cent and basic product with 0.25 per cent share. From the above analysis it can be concluded that in Vijaya dairy milk products are the major product line, followed by by-products. Basic product line is insignificant in terms of its percentage share in total whereas milk products line is significant in product-mix width and sales value.

Table 5.3. Product Line Sales in Vijaya Dairy During 2007

Sl. No.		Milk products	Product Line Width	
			Sales (Rs. lakhs)	%ge Share in Total Sales
1.	(1)	**Milk Products**		
	(a)	Skim Milk	153.46	3.79
	(b)	Double Toned Milk	733.21	18.12
	(c)	Toned Milk	694.62	17.17
	(d)	Standardized Milk	115.77	2.86
	(e)	Gold Milk	2161.04	53.45
		Sub-total	3858.10	95.39
2.	(2)	**Basic Product**		
	(a)	Paneer	10.11	0.25
		Sub-total	10.11	0.25
	(3)	**By Products**		
	(a)	Butter	59.45	1.47
	(b)	Butter Milk	28.05	0.69
	(c)	Ghee Product	33.69	0.84
	(d)	Skimmed milk product	33.61	0.83
	(e)	Doodh peda	2.57	0.06
	(f)	Sterilized flavoured milk	19.10	0.47
		Sub-total	176.41	4.36
		Total (sub-total (1)+(2)+(3)	4044.68	100.00

Source: Primary Data.

Product line width and length of Nandi Dairy is shown for the year 2006-07 in Table 5.4. In terms of sales value and percentage shares in total sales volume milk products accounted for Rs.3151.20 lakhs, by-products Rs. 47.00 lakhs and sweet product Rs.428.94 lakh in aggregative sales volume of Rs. 3627.14 lakhs. In percentage terms milk

Table 5.4. Product Line Sales in Nandi Dairy During 2007

Sl. No.	Milk products	Product Line Width	
		Sales (Rs. lakhs)	%ge Share in Total Sales
1	2	3	4
1.	(1) **Milk Products**		
	(a) Skim Milk	188.20	5.18
	(b) Double Toned Milk	627.80	17.30
	(c) Toned Milk	345.20	6.51
	(d) Standardized Milk	169I.40	4.67
	(e) Whole Milk	1820.60	50.19
	Sub Total	3151.20	86.87
2.	(2) **By Products**		
	(a) Ghee	20.80	0.57
	(b) Butter-milk	26.20	0.72
	Sub Total	47.00	1.29
3.	(3) **Sweet Product**		
	(a) Khoya	65.60	1.80
	(b) *Kalakan*	77.78	2.14
	(c) *Doodh peda*	65.95	1.81
	(d) Milk *Mysur Pack*	76.41	2.10
	(e) *Palakova*	65.76	1.81
	(f) *Makhan*	77.44	2.13
	Sub Total	428.94	11.83
	Total (sub-total ((1)+(2)+(3)	3627.14	100.00

Source: Primary Data.

product line for Nandi dairy was the major contributor to total sales volume with percentage share of 86.87 per cent, followed by sweet products with percentage share of 11.83 per cent, by-products were insignificant with percentage share of 1.29 per cent. Based on the preceding analysis, it can be concluded that milk product line is the major contributed to sales volume followed by sweet products line. Unlike in Vijaya Dairy the share of by-products in total sales volume was very insignificant. Milk product line which is a major product line in Nandi dairy was totally non-existent in Vijaya Dairy. Hence this two dairy sample units differ a lot in terms of product mix and focus on production mix, Vijaya and Nandi dairy focus more on milk product line than on others product lines.

Product Depths in Samples Dairies

Product depth refers to sum total of products in all the product lines put together. High depth implies offering of wide variety of products to the consumers and *vice versa*. In Table 5.5 is shown increasing product depth of Vijaya dairy. To start with this doing had two products in 1977 which number increased to 5 in 1983, 6 in 1987, 7 in 1990, 8 in 1992, 11 in 1998 and 12 in 2001. thus, over 15-year period the number of products offered by this doing increased from 2 in the initial year to 12 in the terminal year, registering a six-fold increase. From the above analysis it can be concluded that the in terms of product d™epth Vijaya dairy has achieved impressive growth.

Table 5.6 presents increasing product depth of Nandi Dairy during the reference period, from 1997 to 2000. Paying attention to the last few of the tables, it can be seen that the number of products offered by Nandi Dairy was 3 in 1997 which number increased to 13 products in 2000, regarding four-fold increase. Combined regarding this table and preceding table it can be concluded the product depth of both dairies are almost same, except on one cannot that the Nandi dairy increased its products depth within a short span of

Table 5.5. Increasing Product Depth of Vijaya Dairy: from 1997 to 2000

Sl. No.	1977	1983	1987	1990	1992	1998	2001
	Description of Individual Products	**Description of Individual Products**	**Description of Individual Products**	**Description of Individual Products**	**Description of Individual Products**	**Description of Individual Products**	**Description of Individual Products**
1	2	3	4	5	6	7	8
1	Gold milk	Gold milk	Gold milk	Gold milk	Gold milk	Gold milk	Gold milk
2	Toned milk	Toned milk	Toned milk	Toned milk	Toned milk	Toned milkT	Toned milk
3		Double toned milk	Double toned milk	Double toned milk	Double toned milk	Double toned milk	Double toned milk
4		*Ghee*	*Ghee*	*Ghee*	*Ghee*	Skimmed milk	Skimmed milk
5		Butter	Butter	Butter	Butter	Standerdised milk	Standerdised milk
6			*Doodh peda*	*Doodh peda*	*Doodh peda*	*Ghee*	*Ghee*
7				Skimmed milk powder	Skimmed milk powder	Butter	Butter
8					Butter milk	*Doodh peda*	*Doodh peda*
9						Skimmed milk powder	Skimmed milk powder
10						Butter milk	Butter milk
11						Sterilized flavoured milk	Sterilized flavoured milk
12							Paneer
Total products	2	5	6	7	8	11	12

Source: Primary Data.

four years as compared to the span of 25 years in Vijaya Dairy.

Table 5.6. Increasing Product Depth of Nandi Dairy: from 1997 to 2000

Sl. No.	Years		
	1997	1999	2000
	Description of individual products	Description of individual products	Description of individual products
1.	Whole milk	Whole milk	Whole milk
2.	Toned milk	Toned milk	Toned milk
3.	Double toned milk	Double toned milk	Double toned milk
4.		Standerdised milk	Standerdised milk
5.		Ghee	Skimmed milk
6.		Butter milk	Ghee
7.			Butter milk
8.			Khoya
9.			Kalakhan
10.			Doodhpeda
11.			Milk mysur pack
12.			Palakova
13.			Makhan
Total products	3	6	13

Source: Primary Data.

Stretching of Packaging and Labelling in Sample Dairies

Packaging inputs safety to the products. This assumes further importance in highly perishable dairy products.

Table 5.7 is drawn to explain packaging and labelling strategies in Vijaya dairy. Like in Nandi dairy tetra-pack is used for all individual products figuring in milk-product line.

Table 5.7. Packing and Labelling Strategies in Vijaya Dairy

Sl. No.	Product line	Packaging	Brand	Information Labelling	
				Fat Content	SNF Content
1.	**(1) Milk Products**				
	(a) Skimmed Milk	Tetra Pack	Vijaya	Indicated	Indicated
	(b) Double Toned Milk	Tetra Pack	Vijaya	Indicated	Indicated
	(c) Toned Milk	Tetra Pack	Vijaya	Indicated	Indicated
	(d) Standardized Milk	Tetra Pack	Vijaya	Indicated	Indicated
	(e) Whole Milk	Tetra Pack	Vijaya	Indicated	Indicated
2.	**(2) Basic Product**				
	(1) Paneer	Cardboard pack	Vijaya	—	—
3.	**(3) By Products**				
	(a) Butter	Tetra Pack	Vijaya	Indicated	Indicated
	(b) Butter Milk	Tetra Pack	Vijaya	Indicated	Indicated
	(c) Ghee	Tetra Pack	Vijaya	Indicated	Indicated
	(d) Skimmed milk Powder	Tetra Pack	Vijaya	Indicated	Indicated
	(e) Doodhpeda	Cardboard pack	Vijaya	No	No
	(f) Sterilized flavoured milk				

Source: Primary Data.

Labelling of individual milk products includes brand name, that is, Vijaya and other information relating to fat and SNF content. Tetra-pack is also used for by-products regarding to labelling Vijaya brand name is used, giving further information such as fat and snf contents for by-products. Namely butter, butter milk, ghee, skimmed milk powder, doodhpeda and sterilized flavoured milk. Apart from brand name, no further information is provided with regarded to by-products namely pannier, skimmed milk powder and doodhpeda, Card board pack is used for paneer which is the only one basic product. Coming to basic product line, as far as labelling is concerned brand name is indicated on cardboard pack and no further information is provided on packing. Like in Nandi Dairy on the packing of individual products of different products shelf life is not shown which is the basic information needed for the consumers.

Packaging and labelling strategies in Nandi dairy are shown in Table 5.8. With respect to milk products and by-products, tetra-pack is used. Labelling includes brand name that is Nandi for all the products. With regard to milk products and by-products labelling additionally covers information relating to fat content and snf content. With regard to sweet products are cardboard pack is used. As far as labelling is concerned only brand name appears with no other information, Shelf life is not indicated either on tetra pack on cardboard pack, which is essential information needed for milk products consumer.

Table 5.8. Packing and Labelling Strategies in Nandi Dairy

S. No.	Product line	Packaging	Brand	Information Labelling	
				Fat Content	SNF Content
1.	**(1) Milk Products**				
	(a) Skimmed Milk	Tetra Pack	Nandi	Indicated	Indicated
	(b) Double Toned Milk	Tetra Pack	Nandi	Indicated	Indicated
	(c) Toned Milk	Tetra Pack	Nandi	Indicated	Indicated
	(d) Standardized Milk	Tetra Pack	Nandi	Indicated	Indicated
	(e) Gold Milk	Tetra Pack	Nandi	Indicated	Indicated
2.	**(2) By Product**				
	(1) Ghee	Tetra Pack	Nandi	Indicated	Indicated
	(2) Butter Milk	Tetra Pack	Nandi	Indicated	Indicated
3.	**(3) Sweet Product**				
	(a) Khawvva	Cardboard pack	Nandi	No	No
	(b) Kalakan	Cardboard pack	Nandi	No	No
	(c) Doodh peda	Cardboard pack	Nandi	No	No
	(d) Milk Mysur Pack	Cardboard pack	Nandi	No	No
	(e) Palakova	Cardboard pack	Nandi	No	No
	(f) Makhan	Cardboard pack	Nandi	No	No

Source: Primary Data.

Ingredient Branding

Ingredient branding is quite informative consumer print of views as it gives details of fat and SNF contains and self life.

Table 5.9 exhibits ingredient branding of milk products in Vijaya dairy for the individual products under each of three product lines. Ingredients such as fat and SNF are shown on packages. Ingredients are indicated in tetra-pack for milk products. The fat content is least at 0.1 per cent for skimmed milk and highest at 6 per cent for gold milk. Coming to SNF it is least in toned milk at 8.5 per cent and highest at

Table 5.9. Ingredient Branding of milk Products in Vijaya Dairy

Sl. No.	Product line	Ingredients		Shelf Line
		Fat	SNF	
	(1) Milk Products			
	(a) Skim Milk	0.1%	8.7%	One day at shelf life
	(b) Double Toned Milk	1.5%	9%	One day at shelf life
	(c) Toned Milk	3%	8.5%	One day at shelf life
	(d) Standardised Milk	4.5%	8.7%	One day at shelf life
	(e) Gold Milk	6%	9%	One day at shelf life
	(2) Basic Product			
	(1) Paneer	25%	30%	1 month
	(3) By-Products			
	(a) Ghee Product	99%	—	6 month
	(b) Butter	84%	2%	3 month
	(c) Butter Milk	1.4%	4%	1 day
	(d) Sterelized Flavoured milk	1.5%	9%	six months
	(e) DOODH PEDA	24%	35%	5 days
	(f) Skimmed milk Powder	1.1%	96%	1 year

Source: Primary Data.

9 per cent in gold milk shelf life is another important feature of milk products, which is uniform with one day shelf life for all milk products. Paneer is the single product coming under basic product line. It has 25 per cent fat, 30 per cent of SNF and one month days shelf life. Compared to milk products pannier is a premium product having more fat percentage and SNF percentage. Hence, it commands premium price. Focusing on by product line, skimmed milk powder has the lowest fat content at one per cent and ghee product highest fat content at 99 per cent. SNF is the lowest at two per cent in butter and highest at 96 per cent in skimmed milk powder. As for as shelf life is concerned it is a five days shelf life for doodhpeda, one day for butter-milk, six months for sterilized flavoured milk, 3 month for butter, 6 months for ghee and 1 year for skimmed powder. Thus, skimmed milk powder, sterilized flavoured milk and ghee have long shelf lives whereas doodhpeda and butter milk have very short shelf life.

Table 5.10 sets out ingredient branding aspects of Nandi Dairy concentrating on milk products it can be observed that fat content is least at 0.1 per cent in skimmed milk and highest at 6 per cent in whole milk. SNF percentage is least at 8.5 per cent in toned milk and highest at 9 per cent and whole milk and double toned milk. Common shelf life is one day for all milk products. Coming to by-products butter milk has the lowest fat content at 1.4 per cent and snf at 4 per cent and short shelf life of 1 day. In contrast ghee has the highest fat content at 99 per cent and longest shelf life of six months. With regard to sweet products all individual products have same features with fat content of 26 per cent snf content of 38 per cent. It is obvious that most of the products of Nandi dairy have very short shelf life, because the major product lines are a sweet product and milk product lines.

Table 5.10. Ingredient Branding of Milk Products in Nandi Dairy

Sl. No.	Product line	Ingredients		Shelf Line
		Fat	SNF	
1.	**(1) Milk Products**			
	(a) Skim Milk	0.1%	8.7%	One day at shelf life
	(b) Double Toned Milk	1.5%	9%	One day at shelf life
	(c) Toned Milk	3%	8.5%	One day at shelf life
	(d) Standardised Milk	4.5%	8.7%	One day at shelf life
	(e) Whole Milk	6%	9%	One day at shelf life
2.	**(2) By Products**			
	(1) Ghee	99%	—	6 months
	(2) Butter milk	1.4	4%	1 days
3.	**(3) Sweet Products**			
	(a) Khavva	26%	38%	5 days
	(b) Kalakan	26%	38%	5 days
	(c) Doodh peda	26%	38%	5 days
	(d) Milk Mysur Pack	26%	38%	5 days
	(e) Palakova	26%	38%	5 days
	(f) Makhan	26%	38%	5 days

Source: Primary data.

Up and Down-Stretching of Product Lines

Up-stretching of product line aims at reaching premium markets where as it's opposite, down-stretching, caters to mass markets. Both intend to expand market to the products in the respective product lines.

Product line stretching in Nandi dairy is presented in Table 5.11. Stretching practices to lengthen the product line are employed in Nandi Dairy with regard to milk products. It offers skimmed milk at Rs.15 per litre in the market. It practices up market stretch offering double toned milk at Rs.17 per litre, toned milk at Rs.18 per litre and standardize

Table 5.11. Product Line and Stretching in Nandi Dairy in 2007

Sl. No	Product line/ item	Up-market Stretch (Sales, Rs. in lakhs)	Down-market Stretch	No Stretch (Sales, Rs. in lakhs)	Sales (Rs. in lakhs)
1.	**(1) Milk Products**				
	(a) Skim Milk		...	188.20	...
	(b) Double Toned Milk	627.80 (Rs. 17.00 1000 ml)	...		
	(c) Toned Milk	345.20 (Rs. 18.00 (1000 ml)	...		
	(d) Standardized Milk	169.40 (Rs. 20.00 (1000 ml)	...		
	(e) Whole Milk	1820.60 (Rs. 22.00 1000 ml)	...		
2.	**(2) By-products**				
	(a) Ghee		...	20.80 (Rs. 160 1000 g)	
	(b) Butter-milk		...	20.80 (Rs. 160 1000 g)	
3.	**(3) Sweet Product**				
	(a) Khawvva		...	65.60 (Rs. 120.00 1000 g)	
	(b) Kalakan		...	77.78 (Rs. 120.00 1000 g)	
	(c) Doodh peda		...	65.95 (Rs. 120.00 1000 g)	
	(d) Milk Mysur Pack		...	76.41 (Rs. 120.00 1000 g)	
	(e) Palakova		...	65.76 (Rs. 120.00 1000 g)	
	(f) Makhan		...	77.44 (Rs. 120.00 1000 g)	
	Total (Rs. in lakhs (3+5)	2963.00 (81.68)	...	664.14 (18.32)	3627.14 (100.00)

Note: Figures in parentheses are price rates of individual products.
Source: Primary Data.

milk at Rs.20 per litre and whole milk at Rs. 22 per litre. With regard to by-products there is no stretch, which is neither up-market stretch, nor down market stretch. *Ghee* is offered at Rs.160 per kg and butter milk is sold at Rs. 4 for 200 ml, with regard to sweet products also no stretching practices are noticed. All sweet products are sold at Rs.120 per kg. There is ample room for Nandi dairies 81.68 per cent of sales revenue from up-market stretch and the remain 18.32 per cent from down-market stretch practising up-market stretch and down-market stretch with regard to by-products line and sweet product line in Nandi market.

Product line stretching practices to lengthen product line in Vijaya Dairy are displayed in Table 5.12. Like Nandi Dairy and Vijaya Doffers the skimmed milk at Rs.15 per litre. Practicing up market stretch, it offers milk products such as double toned milk at Rs. 17 per litre, toned milk at Rs.18 per litre and whole milk at Rs.22 per litre Vijay dairy whereas 91.59 per cent of its total sales revenue from up-market stretch and the set of 8.41 per cent from no-stretch. Down-stretch market stretch is not at we practiced is this dairy like in Nandi dairy. There is ample scope to adopt stretching practices with respect to basic product, that is, pannier and by-products.

Trend of Sales of Core and Staple Dairy Products

Alternating dairy products can we classified in to core and stable products. The sales trend of these two product categories are discussed in the following lines. Table 5.13 is presents yearly sales data of products reclassified as core products and staple products of Vijaya dairy for 10-year period from 1998 to 2007. The reclassification of products results in categorizing products into core products and staple products. As can be seen from column thirteen core products accounted for 90.40 per cent total sales whereas staple products for 9.60 per cent of total sales. Under core products the LGRs are not significant at 5 per cent level for two out of

Table 5.12. Product Line and Stretching in Vijaya Dairy in 2007

Sl. No	Product line/ item	Up-market Stretch (Sales, Rs. in lakhs)	Down-market Stretch	No Stretch (Sales, Rs. in lakhs)	Sales (Rs. in lakhs)
1.	**(1) Milk Products**				
	(a) Skim Milk		...	153.46 (Rs. 15.00 1000 ml)	...
	(b) Double Toned Milk	733.21 (Rs. 17 1000 ml)	...		
	(c) Toned Milk	694.62 (Rs. 18 (1000 ml)	...		
	(d) Standardized Milk	115.77 (Rs. 20 (1000 ml)	...		
	(e) Gold Milk	2161.04 (Rs. 22 1000 ml)	...		
2.	**(2) Basic products**				
	(a) Pannir		...		
3.	**(3) By Product**				
	(a) Butter		...	59.45 (Rs. 140 1000 g)	
	(b) Butter Milk		...	28.05 (Rs. 4.00.00 200 ml)	
	(c) Ghee Product		...	33.69 (Rs. 160 1000 g)	
	(d) Skimmed milk product		...	33.61 (Rs. 140 1000 g)	
	(e) Doodh peda		...	2.57 (Rs. 120 1000 g)	
	(f) Sterilized flavoured milk		...	19.10 (Rs. 10 200 ml)	
	Total (Rs. in lakhs (3+5)	3704.64 (91.59)	...	240.04 (8.41)	4044.68 (100.00)

Note: Figures in parentheses are price rates of individual products.
Source: Primary Data.

Table 5.13. Trend of Sales of Core and Staple Products of Vijaya Dairy: From 1998 to 2007

Sl. No.	Product Classification	Yearly Sales (in Rs. lakhs)										Average Sales (in Rs. lakhs)	LGR	/t/ value	r. value
		1998	1999	2000	2001	2002	2003	2004	2005	2006	2007				
1	2	3	4	5	6	7	8	9	10	11	12	13	14	15	16
1	**Core Product**														
	Double toned milk	186.15	219.24	191.55	240.56	484.44	305.82	423.51	445.4	694.98	733.21	392.49 (16.49)	15.27	5.96 **	0.90
	Toned milk	210.97	243.6	255.61	267.29	635.25	339.8	468.74	471.6	772.2	694.62	435.97 (18.32)	13.12	4.44 **	0.84
	Gold milk	683.1	609	689.58	668.23	1633.94	883.86	1159	1467.87	2123.75	2161.04	1207.84 (50.77)	14.31	4.87 **	0.86
	Ghee	42.67	42.44	45.24	40.53	38.19	24.38	36.58	32.2	31.4	33.69	36.73 (1.54)	-4.10	2.89 **	-0.71
	Butter	48.75	59.53	72.84	35.19	46.06	51.82	58.44	46.79	13.16	59.45	49.20 (2.06)	-3.48	0.95 NS	-0.31
	Skimmed milk powder	15.59	15.18	60.38	9.37	14.87	30.6	33.84	37.61	32.53	33.61	28.36 (1.19)	5.53	0.93 NS	0.31
	Sub-total											2150.59 (90.40)			

1	2	3	4	5	6	7	8	9	10	11	12	13	14	15	16
2	Staple Product														
	Skim milk	62.05	12.21	76.62	66.82	151.25	84.95	89.32	131	154.44	154.36	98.30 (4.13)	12.94	3.83 **	0.80
	Standerdised milk	99.28	60.9	63.85	93.55	121	84.95	89	104.8	115.83	115.77	94.90 (3.98)	4.40	2.14 NS	0.60
	Pannir	...	...	...	0.67	0.98	2.27	3.3	8.31	9.79	10.11	5.06 (0.21)	36.67	7.43 **	0.96
	Doodh peda	1.04	0.93	0.97	1.11	1.45	1.64	1.54	1.89	2.1	2.57	1.52 (0.06)	11.15	0.66 NS	0.23
	Sterilized flavoured milk	17.72	18.15	17.45	2.23	19.12	22.84	23.27	21.1	16.39	19.1	17.74 (0.74)	2.91	0.90 NS	0.30
	Butter milk	3.51	4.69	4.27	6.51	7.6	12.38	9.19	12.76	20.62	28.05	10.96 (0.46)	21.43	5.78 **	0.90
3	Sub-total											228.48 (9.60)			
	Total (1+2)											2379.07 (100.00)			

Note: ** Significant at 1 % level, NS: Not significant at 5 % level.

Source: Primary Data.

six products; for another three products LGRs are significant at 1 per cent level and for one products only LGRs significant at 1 per cent level. With regards to staple products LGRs are not significant for three out of six products; and for other three products LGRs are significant at one per cent level.

Table 5.14 show the sales of core and staple products of Nandi dairy for 10-year period from 1998-2007. Reading column 13 it can be seen that core products accounted for 83.43 per cent if total sales, whereas by-products for 16.57 per cent of total sales. The LGRs of all products registered under core product and staple products categories are significant at one per cent level. Combinable reading Tables 5.13 and 5.12, it can be concluded that LGRs of Nandi products under core product and staple product categories have registered LGRs which are significant at one per cent level, as compare to LGRs of products of Nandi dairy which are mostly non significant at five per cent level.

Distribution of Sales By-product Lines

This section presents the distribution of sales by product lines in the samples dairies. Table 5.15 displays product-wise distribution of sales under different product lines of Vijaya Dairy for ten year period from 1998 to 2007. Focusing on column 13 it can be said that milk-product line is the major product line accounting for 93.71 per cent in total sales, followed by by-product line with 6 per cent and basic product line 0.21 per cent of total sales. LGRs are significant at one per cent level for milk product line except double toned milk and products of other product lines. With regard to products of by products line namely butter, skimmed milk powder *doodhpeda*, sterilized flavoured milk; LGRs are not significant at five per cent level. With respective to ghee LGR is significant at five per cent level. For the rest of products LGRs are significant at one per cent level. With regard to one product, coming under basic product namely pannier LGR is significant at one per cent level.

Table 5.14. Trends of Sales of Core and Staple Products of Nandi Dairy: From 1998 to 2007 (Rs. in lakhs)

Sl. No.	Product Classification	Yearly Sales (in Rs. lakhs)										Average Sales (in Rs. lakhs)	LGR	/t/ value	r. value
		1998	1999	2000	2001	2002	2003	2004	2005	2006	2007				
1	2	3	4	5	6	7	8	9	10	11	12	13	14	15	16
1	**Core Product**														
	Double toned milk	183.84	193.12	285.43	344.88	324.54	395.20	467.20	512.46	554.80	627.80	388.93 (17.69)	12.62	19.07 **	0.99
	Toned milk	87.52	70.24	111.69	180.54	187.74	208.05	256.8	281.77	277.4	345.2	200.70 (9.12)	14.70	13.77 **	0.98
	Whole milk	525.6	535.68	719.78	952.56	1007.82	1227.4	1354.8	1538.16	1636.6	1820.6	1131.90 (51.47	13.32	29.18 **	0.99
	Ghee	9.6	10.88	11.36	11.84	12	12.96	13.28	14.24	15.36	20.8	13.23 (0.60)	6.95	5.47 **	0.88
	Butter milk	9.35	9.55	12.2	17.5	17.81	19.5	18.64	21.45	20.06	26.2	17.23 (0.78)	9.73	8.08 **	0.94
	Kalakhan	...	...	16.14	31.45	37.85	34.98	40.7	43.32	40.9	77.78	40.39 (1.83)	14.76	3.80 **	0.84
	Makhan	...	...	17.6	27.6	34.15	37.69	43.42	50.77	49.45	77.44	42.30 (1.92)	16.44	6.10 **	0.94
	Sub-total											1834.68 (83.43)			

1	2	3	4	5	6	7	8	9	10	11	12	13	14	15	16
2	**Staple Products**														
	Skimmed milk	35.36	35.04	74.46	82.06	102.42	145.54	116.8	128.11	166.4	188.2	107.40 (4.88)	15.36	9.70 **	0.96
	Standardised milk	43.68	43.84	49.46	82.46	85.32	103.93	140.00	102.37	138.60	169.40	95.90 (4.36)	14.22	8.23 **	0.94
	Khoya	...	...	18.21	28.54	34.45	36.67	41.1	45.85	43.81	65.6	39.28 (1.78)	13.53	6.64 **	0.93
	Doodh peda	...	...	16.45	29.64	36.6	38.34	39.54	45.08	44.94	65.95	39.57 (1.79)	13.52	6.08 **	0.93
	Milk mysur pack	...	...	19.6	30.15	34.35	36.22	42.58	47.3	46.93	76.41	41.69 (1.89)	15.04	5.58 **	0.91
	Palakova	...	...	18.45	31.1	35.4	35.98	41.46	47.98	48.24	65.76	40.54 (1.84)	13.50	8.08 **	0.95
	Sub-total											364.38 (16.57)			
	Total (2)											2199.06 (100.00)			

Note: **Significant at 1 % level.
Source: Primary Data.

Table 5.15. Product Line Distribution of Vijaya Dairy Sales: From 1998 to 2007

(Rs. in lakhs)

		Years													
Sl. No.	Product line	1998	1999	2000	2001	2002	2003	2004	2005	2006	2007	Avg.	LGR	/t/ value	r. value
1	2	3	4	5	6	7	8	9	10	11	12	13	14	15	16
1	**Milk Products**														
	Skimmed milk	62.05	12.21	76.62	66.82	151.25	84.95	89.32	131.00	154.44	154.36	98.30 (4.15)	12.94	3.83 **	0.80
	Double toned milk	186.15	219.24	191.55	240.56	484.44	305.82	423.51	445.40	694.98	733.21	392.49 (16.57)	15.27	5.96 **	0.90
	Toned milk	210.97	243.60	255.61	267.29	635.25	339.80	468.74	471.60	772.20	694.62	435.97 (18.40)	13.12	4.44 **	0.84
	Standardised milk	99.28	60.90	63.85	93.55	121.00	84.95	89.00	104.80	115.83	115.77	94.90 (4.00)	4.40	2.14 NS	0.60
	Gold milk	683.10	609.00	689.58	668.23	1633.94	883.86	1159.00	1467.87	2123.75	2161.04	1207.84 (51.00)	14.31	4.87 **	0.86
	Sub-total	1241.92	1218.03	1277.21	1336.47	3025.44	1699.38	2229.65	2620.67	3861.20	3859.00	2229. 50 (93.71)	13.58	4.97 **	0.87
2	**Basic Product**														
	Pannir	...	...	...	0.67	0.67	2.27	3.30	8.31	9.79	10.11	5.06 (0.21)	36.67	7.43 **	0.96
	Sub-total	...	...	...	0.67	0.67	2.27	3.30	8.31	9.79	10.11	5.06 (0.21)	36.67	7.43 **	0.96

1	2	3	4	5	6	7	8	9	10	11	12	13	14	15	16
3	**By-products**														
	Butter	48.75	59.53	72.84	35.19	46.06	51.82	58.44	46.79	13.16	59.45	49.20 (2.07)	-3.48	0.95 NS	-0.31
	Butter Milk	3.51	4.69	4.27	6.51	7.60	12.38	9.19	12.76	20.62	28.05	10.96 (0.46)	21.43	5.78 **	0.90
	Ghee	42.67	42.44	45.24	40.53	38.19	24.38	36.58	32.20	31.40	33.69	36.73 (1.55)	-4.10	2.89 *	-0.71
	Skimmed milk powder	15.59	15.18	60.38	9.37	14.87	30.60	33.84	37.61	32.53	33.61	28.36 (1.20)	5.53	0.93 NS	0.31
	Doodh peda	1.04	0.93	0.97	1.11	1.45	1.64	1.54	1.89	2.10	2.57	1.52 (0.6)	11.15	0.66 NS	0.23
	Sterilized flavoured milk	17.72	18.15	17.45	2.23	19.12	22.84	23.27	21.10	16.39	19.10	17.74 (0.74)	2.91	0.90 NS	0.30
	Sub-total	129.30	140.92	146.82	95.62	128.28	145.97	166.18	160.68	126.01	186.60	144.51 (6.02)	3.17	1.94 NS	0.56
	Total (1+2+3)	1371.23	1358.96	1424.04	1432.10	3153.73	1845.35	2395.83	2781.35	3873.81	4044.68	2379.07 (100.00)	12.806	.10 **	0.90

Note: ** Significant at 1 % level,
* Significant at 5 % level, NS: Not significant at 5 % level.

Source: Primary Data.

Table: 5.16. Product Line Distribution of Nandi Dairy Sales: From 1998 to 2007

(Rs. in lakhs)

		Years													
Sl. No.	Product line	1998	1999	2000	2001	2002	2003	2004	2005	2006	2007	Avg.	LGR	/t/ value	r. value
1	2	3	4	5	6	7	8	9	10	11	12	13	14	15	16
1	**Milk Products**														
	Skimmed milk	35.36	35.04	74.46	82.06	102.42	145.54	116.80	128.60	128.11	188.20	107.40 (5.00)	15.36	9.70 **	0.96
	Standerdized milk	43.68	43.84	49.46	82.46	85.32	103.93	140.00	101.50	102.37	169.40	95.90 (4.46)	14.22	8.23 **	0.94
	Toned milk	87.52	70.24	111.69	180.54	187.74	208.05	256.80	272.35	281.77	345.20	200.70 (9.34)	14.70	13.77 **	0.98
	Double toned milk	183.84	193.12	285.43	344.88	324.54	395.20	467.20	507.80	512.46	627.80	388.93 (18.10)	12.62	19.07 **	0.99
	Whole milk	525.60	535.68	719.78	952.56	1007.82	1227.40	1354.80	1532.76	1538.16	1820.60	1131.90 (52.69)	13.32	29.18 **	0.99
	Sub-total	876.00	877.92	1241.00	1642.50	1708.00	2080.12	2335.60	2543.01	2562.87	3151.20	1924.44 (87.51)	13.48	26.81 **	0.99
2	**By-products**														
	Ghee	9.60	10.88	11.36	11.84	12.00	12.96	13.28	15.06	15.36	20.80	13.23 (0.61)	6.95	5.47 **	0.88

1	2	3	4	5	6	7	8	9	10	11	12	13	14	15	16
	Butter milk	9.35	9.55	12.20	17.50	17.81	19.50	18.64	19.70	20.06	26.20	17.23 (0.62)	9.73	8.08 **	0.94
	Sub-total	18.95	20.43	23.56	29.34	29.81	32.46	31.92	34.76	35.42	47.00	30.46 (1.38)	8.52	8.77 **	0.95
3	**Sweet Products**														
	Khoya	....	....	18.21	28.54	34.45	36.67	41.10	45.85	43.81	65.60	39.28 (1.82)	13.53	6.64 **	0.93
	Kalakan	....	....	16.14	31.45	37.85	34.98	40.70	43.32	40.90	77.78	40.39 (1.88)	14.76	3.80 **	0.84
	Doodh peda	....	....	16.45	29.64	36.60	38.34	39.54	45.08	44.94	65.95	39.57 (1.84)	13.52	6.08 **	0.93
	Milk mysur pack	....	.....	19.60	30.15	34.35	36.22	42.58	47.30	46.93	76.41	41.69 (1.94)	15.04	5.58 **	0.91
	Palakova	....	....	18.45	31.10	35.40	35.98	41.46	47.98	48.24	65.76	40.54 (1.88)	13.50	8.08 **	0.95
	Makhan	....	.....	17.60	27.60	34.15	37.69	43.42	50.77	49.45	77.44	42.30 (1.96)	16.44	6.10 **	0.94
	sub-total	....	....	106.45	178.48	212.80	219.88	248.80	280.33	274.27	428.94	243.74 (11.08)	14.49	5.98 **	0.92
	Total(1+2+3)	8.94.95	898.35	1347.45	1850.32	1950.61	2332.46	2616.32	2858.10	3083.49	3627.14	2199.00 (100.0)	14.17	23.86 **	0.99

Note: ** Significant at 1 % level.
Source: Primary Data.

Table 5.16 exhibits product line sales and sales of individual products under each product line year-wise for Nandi dairy for reference period from 1998 to 2007. Focusing on column thirteen, it can be observed that milk product line is the major product line with 89.57 per cent in total sales, followed by sweet-product line with 9.35 and by-product line with 1.41 per cent. For all individual products under three product lines the LGRs are significant at 1 per cent level. LGRs of sales of individual products have high level significant in Nandi dairy as compare to those of Vijaya dairy.

Conclusion

Vijaya and Nandi dairies have three product lines each. Dairy product form five-layer product hierarchy in each of the dairies milk product line is dominant in terms of sales while the other two being in significant. Ever increasing trend can be noticed with regarding to product depth. Both the dairies have attractive packaging for their products which are sold under popular brands. Ingredient branding is adopted in both the dairies, which specifies fat and SNF contents. Both up- and down-stretching practices are adopted to cater to premium and mass markets.

6 CHAPTER

PRICING OBJECTIVES AND METHODS OF DAIRIES

INTRODUCTION

In sequence of order of four P's conceptual framework of marketing second 'P' refers to price. Price-related marketing practices form part of this chapter. Pricing objectives, demand estimation methods, pricing methods, price discounts and allowances are discussed in the following sections of the chapter.

Pricing Objectives

Pricing objectives not a single pricing objective pursued. Multiple are the objectives of pricing the dairy products. Their section deals with pricing objectives. In Table 6.1 are shown the pricing objectives corresponding to individual dairy products of Vijaya dairy unit. Altogether four pricing objectives are pursued namely survival maximum profit, maximum products market share and product quality leadership. Survival objective is the pricing objective with respective to skimmed milk maximum profit pricing objectives is followed with respect to eight products such as butter, buttermilk *ghee*, toned milk, *dodhpeda*, etc. Market share objective is pursued with respect to pannier and standardised milk, and product quality leadership is pricing objective with regard to whole milk. From the preceding analysis it is obvious that maximum profit objective is followed with regard is eight of products of Vijaya dairy followed by market share for two products and survival and product quality leadership for one product each. Hence, it

Table 6.1. Pricing Objectives of Different Products in Vijaya Dairy

Sl. No.	Pricing objectives	Individual products corresponding	No. of products
1	Survival	Skimmed milk	1
2	Maximum Profit	Double toned milk	8
		Toned milk	
		Sterilized flavoured milk	
		Butter	
		Butter milk	
		Ghee	
		Skimmed milk powder	
		Doodh peda	
3	Market Share	Pannir	2
		Standerdized milk	
4	Product Quality Leadership	Gold milk	1
	Total Products		12

Source: Primary Data

can be concluded maximum profit is the major pricing objective in Vijaya dairy.

Pricing objectives pursued in Nandi dairy are shown in Table 6.2. Design of the table is similar to the preceding table. Like in Vijaya dairy and Nandi dairy also adopts four pricing objectives such as survival, maximum profit market share, and product quality leadership. Major pricing objective is a maximum profit pursued with regard to ten products. In respect of the remaining three objectives one product each comes under the three objectives. Based on the above analysis it can be concluded that maximum profit objective is a major pricing objective for ten out of thirteen products in Nandi dairy.

Table 6.2. Pricing Objectives of Different Products in Nandi Dairy

Sl. No.	Pricing objectives	Corresponding of Individual products	No. of products
1	Survival	Skimmed milk	1
2	Maximum profit	Double toned milk	
		Toned milk	
		Koya	
		Kalakhan	
		Doodh peda	10
		Milk *mysur pak*	
		Palakova	
		Makhan	
		Ghee	
		Butter milk	
3	Market share	Standerdised milk	1
4	Product quality leadership	Whole milk	1
	Total products		13

Source: Primary Data.

Demands Estimations Techniques

Three broad techniques used or demand estimation in two dairies are statistical analysis price experiments and surveys. Table 6.3 illustrates the methods of demand estimation used in Vijaya dairy unit. Three methods are in vogue namely statistical analysis, price experiment and surveys, with Statistical analysis as a method of demand estimation is followed for all twelve products. Price experiments are followed for six products. For another six products survey technique is followed. From the above

Table 6.3. Methods of Demand Estimation of Products in Vijaya Dairy

Sl. No.	Demand estimate methods	Corresponding of dairy products	No. of products
1	Statistical Analysis	(1) Milk Products	12
		(a) Skim Milk	
		(b) Double Toned Milk	
		(c) Toned Milk	
		(d) Standardized Milk	
		(e) Gold Milk	
		(2) Basic Product	
		(a) Pannier	
		(3) By Products	
		(a) Butter	
		(b) Butter Milk	
		(c) Ghee Product	
		(d) Skimmed milk product	
		(e) Doodh peda	
		(f) Sterilized flavoured milk	
2	Price Experiments	(a) Skim Milk	5
		(b) Double Toned Milk	
		(c) Toned Milk	
		(d) Standardized Milk	
		(e) Gold Milk	
3	Surrey's	(a) Skim Milk	5
		(b) Double Toned Milk	
		(c) Toned Milk	
		(d) Standardized Milk	
		(e) Gold Milk	

Source: Primary Data

analysis it can be concluded that statistical analysis is the demand estimation techniques mostly followed with regard to all products. Price experiments and surveys are of equal importance next to statistical analysis, which are pursued with regard to six products each.

Table 6.4. Methods of Demand Estimation of Products in Nandi Dairy

Sl. No.	Demand estimate methods	Corresponding of dairy products	No. of products
1	Statistical Analysis	(1) Milk Products (a) Skim Milk (b) Double Toned Milk (c) Toned Milk (d) Standardized Milk (e) Whole Milk (2) By-products (a) *Ghee* (b) Butter-milk (3) Sweet Product (a) *Khoya* (b) *Kalakan* (c) *Doodh peda* (d) Milk *Mysur Pack* (e) *Palakova* (f) *Makhan*	13
2	Price Experiments	(1) Milk Products (a) Skim Milk (b) Double Toned Milk (c) Toned Milk (d) Standardized Milk (e) Whole Milk	5
3.	Surrey's	(1) Milk Products (a) Skim Milk (b) Double Toned Milk (c) Toned Milk (d) Standardized Milk (e) Whole Milk	5

Source: Primary Data

Table 6.4 is designed on the lines of preceding table, indicating methods of demand estimation corresponding to individual dairy products of Nandi dairy unit. Like in Vijaya dairy statistical analysis is the technique of demand estimation of all 13 products of Nandi dairy followed by price experiments and surreys corresponding to six dairy products each. Based on the above analysis it is oblivious that statistical analysis is popular method of demand estimation of dairy products, since it is used in all thirteen dairy products of Nandi dairy unit.

PRICING METHODS

Pricing methods used in the sample dairies are multiple in numbers and varied in their nature. This section discusses then in detail. Table 6.5 focuses on the pricing methods

Table 6.5. Pricing Methods in Vijaya Dairy

Sl. No.	Pricing methods	Corresponding products of Vijaya	No. of products
1	2	3	4
1	Market Pricing	(1) Milk Products (a) Skim Milk (b) Double Toned Milk (c) Toned Milk (d) Standardized Milk (e) Gold Milk (2) Basic Product (a) Panner (3) By Products (a) Butter (b) Butter Milk (c) Ghee Product (d) Skimmed milk product (e) Doodh peda (f) Sterilized flavoured milk	12
2	Target Return Pricing	(1) Milk Products (a) Skim Milk (b) Double Toned Milk	5

1	2	3	4
		(c) Toned Milk (d) Standardized Milk (e) Gold Milk	
3	Perceived Value Pricing	(1) Milk Products (a) Skim Milk (b) Double Toned Milk (c) Toned Milk (d) Standardized Milk (e) Gold Milk	5
4.	Value Pricing	(1) Milk Products (a) Skim Milk (b) Double Toned Milk (c) Toned Milk (d) Standardized Milk (e) Gold Milk (2) Basic Product (a) Panner (3) By Products (a) Butter (b) Butter Milk (c) Ghee Product (d) Skimmed milk product (e) Doodh peda (f) Sterilized flavoured milk	12
5.	Going Rate Pricing	(1) Milk Products (a) Skim Milk (b) Double Toned Milk (c) Toned Milk (d) Standardized Milk (e) Gold Milk (2) Basic Product (a) Panner (3) By Products (a) Butter (b) Butter Milk (c) Ghee Product (d) Skimmed milk product (e) Doodh peda (f) Sterilized flavoured milk	12

Source: Primary data

followed in Vijaya dairy unit namely market pricing, target pricing, perceived value pricing, value pricing and going rate pricing. Market pricing method is used for all the twelve products of Vijaya dairy unit. Similar is the case with the value pricing and going rate pricing, which are adopted for all the products. Target return pricing and perceived value pricing are adopted. Thus, with regard to five products each market pricing and value pricing and going rate pricing are universal methods of pricing in this dairy units that is for all the products of the dairy units these pricing methods are employed.

Table 6.6 is drawn to highlight the pricing method practised in Nandi dairy unit. The design of this table is akin to the preceding table wherein pricing methods are shown corresponding to the individual products of dairy unit. Market pricing, value pricing and going rate pricing rate methods are universally employed with regard to all products of the dairy unit. Target return pricing and perceived value pricing are adopted with regard to five products each. Comparing Table 6.5 and 6.6 it can be noticed the pricing methods adopted in the two sample dairy units are almost same.

Table 6.6. Pricing Methods in Nandi Dairy

Sl. No.	Pricing methods	Corresponding products of Vijaya	No. of products
1	2	3	4
1	Market Pricing	(1) Milk Products (a) Skim Milk (b) Double Toned Milk (c) Toned Milk (d) Standardized Milk (e) Whole Milk (2) By-products (a) Ghee (b) Butter-milk (3) Sweet Products (a) *Khoya* (b) *Kalakan* (c) *Doodh peda* (d) Milk *Mysur Pak* (e) *Palakova* (f) *Makhan*	13

1	2	3	4
2	Target Return Pricing	(1) Milk Products (a) Skim Milk (b) Double Toned Milk (c) Toned Milk (d) Standardized Milk (e) Whole Milk	5
3	Perceived Value Pricing	(1) Milk Products (a) Skim Milk (b) Double Toned Milk (c) Toned Milk (d) Standardized Milk (e) Whole Milk	5
4.	Value Pricing	(1) Milk Products (a) Skim Milk (b) Double Toned Milk (c) Toned Milk (d) Standardized Milk (e) Whole Milk (2) By-products (a) *Ghee* (b) Butter-milk (3) Sweet Products (a) *Khoya* (b) *Kalakan* (c) *Doodh peda* (d) Milk *Mysur Pack* (e) Palakova (f) *Makhan*	13
5.	Going Rate Pricing	(1) Milk Products (a) Skim Milk (b) Double Toned Milk (c) Toned Milk (d) Standardized Milk (e) WholeMilk (2) By-products (a) Ghee (b) Butter-milk (3) Sweet Products (a) *Khoya* (b) Kalakan (c) *Doodh peda* (d) Milk *Mysur Pack* (e) *Palakova* (f) *Makhan*	13

Source: Primary Data

Price Discount and Allowances

To promote sales of dairy products, sample units offer price discounts and allowances. Price discounts and allowances offered in Vijaya and Nandi dairy units are shown in Table 6.7. As can be seen from the column Two of the table there are five types of discounts and allowances offered to dealers and consumers. To be more special discounts and allowances are cash discount, quality discount, function discount, seasonal discount and allowance. Both dairies offer cash discount. Both of then offer higher cash discount, ten per cent, for dealers followed by eight per cent to institutional, consumers. With regard to non-institutions consumers there is a marginal lower cash discounts are offered by two diaries. Nandi dairy offering seven per cent discount which is one per cent higher than six per cent offered by Vijaya dairy unit quantity discounts are offered by both the dairies for dealers of the same magnitude, that is, ten per cent in each case. With regard to institutional consumers Nandi dairy offers higher quantity discounts at eight per cent whereas Vijaya dairy offers quantity discounts of seven per cent. The practice of offering seasonal discounts are also noticed in the two sample dairy units of Vijaya and Nandi offering ten per cent seasonal discount for dealers whereas Nandi dairy offering seven per cent seasonal discount the institutional consumers and Vijaya six per cent. Both the dairies offer five per cent seasonal discount to non-institutional to this seven per cent of consumers. Allowances are also offered to the channel members to promote sales. With regard to the allowances both sample dairy units offer eight per cent allowances to their dealers.

Table: 6.7. Price Discounts and Allowances in Vijaya and Nandi Dairies

Sl. No.	Description	Dealers	Consumers	
			Institutional	Non-institutional
1	Cash Discounts a) Vijaya b) Nandi	 10% 10%	 8% 8%	 6% 7%
2	Quantity Discount (a) Vijaya (b) Nandi	 10% 10%	 7% 8%	 NA NA
3	Functional Discount (a) Vijaya (b) Nandi	 8% 8%	 NA NA	 NA NA
4	Seasonal Discount (a) Vijaya (b) Nandi	 10% 10%	 6% 7%	 5% 5%
5	Allowance Discount (a) Vijaya (b) Nandi			

Source: Primary Data

Conclusion

With record to pricing objectives, profit maximization objective is popular objective pursued for most of the products. Similarly statistical analysis is the widely followed demand estimation technique in both the dairies. Market pricing and value pricing are the most useful pricing methods in these dairies. Price discounts and allowances and offered to dealers and institutional and individual consumers to promote sales.

7 CHAPTER

PROMOTION PRACTICES IN SAMPLE DAIRY UNITS

INTRODUCTION

This chapter deals with practices related to promotion mix of marketing third 'p' furnished to promotion. Accordingly advertisement programs, events/experiences programs, personal selling programs, direct marketing methods used for determining communication budgets, communication budgets outlays, budgets *vis-à-vis* sales, advertising budgets, budgets vis-à-vis sales, sales promotion tools and media-wise advertising are analyzed in what fallows.

Advertising Programmes in Sample Dairy Units

Advertising targets the change of consumer mind in favour products of the business concern, since hurdle product market expansion lies in consumer mind. This section is devoted to study advertising practices of the sample dairy units.

Table 7.1 presents qualitative data relating to advertising programmes in Vijaya and Nandi dairy units. Thirteen advertisement programmes are identified. Marketing or contacted personnel are asked to respond to check list items of advertisement programmes. If the programmes are adopted 'tick' mark is shown against that programme and 'cross' mark is used against the programmes is not used. Out of thirteen programmes, thirteen ticks are obtained for Vijaya dairy and twelve for Nandi dairy. The ticks are converted into percentage scores dividing number of ticks/number of programmes and multiplying ratio by hundred. The percentage score of checklist method is hundred for Vijaya dairy 92.30 per cent for Nandi dairy. Based on the single measure of percentage score of checklist it can be concluded that both the dairies followed all advertising programmes their sales promotion.

Table 7.1. Advertising Programmes in Vijaya and Nandi Dairies

Sl. No.	Advertising programmes	Vijaya	Nandi
1	Print and broadcast ads	✔	✔
2	Packaging-outer	✔	✔
3	Motion pictures	✔	✔
4	Broachers and booklets	✔	✔
5	Posters and leaflets	✔	✔
6	Directories	✔	✔
7	Billboards	✔	✔
8	Reprints of ads	✔	✔
9	Display signs	✔	✔
10	Point of purchase displays	✔	✔
11	Audiovisual material	✔	×
12	Symbols and logos	✔	✔
13	Video tapes	✔	✔
	Total ticks	13	12

Legend: *(✔) indicates the presence of the programme.*
(×) Indicates the obscene of the programme

Notes: 1 Percentage score of checklist = $\frac{\text{No. of ticks}}{\text{No. of programmes}} \times 100$

2. Index for Vijaya dairy = 100 %
Index for Nandi dairy = 92.3 %

Source: Compiled from the records of Vijaya and Nandi dairy,

Sales Promotion Programmes

These programs are initiated at dealer places, giving them proper incentives to undertake these programs so as to induce the consumers to purchase company products.

Table 7.2 presents' sales promotion programmes adopted in Vijaya and Nandi dairy units. Thirteen sales promotion programmes generally adopted in the dairy industry are listed in the table. Against these programmes, ticks and cross marks are shown indicating presence and allowance of programmes respectively. Out of thirteen sales promotion programmes there are four programmes adopted in Vijaya dairy unit and eight in Nandi dairy unit.

Table 7.2. Sales Promotion Programmes in Vijaya and Nandi Dairies

Sl. No.	Sales promotion	Vijaya	Nandi
1	Contests, games Sweepstakes, lotteries	×	✔
2	Premiums and gifts	×	✔
3	Sampling	×	✔
4	Fairs and trade shows	✔	✔
5	Exhibits	✔	✔
6	Demonstrations	✔	✔
7	Coupons	×	✔
8	Rebates	×	×
9	Low-interest financing	×	×
10	Entertainment	×	×
11	Trade-in allowances	✔	✔
12	Continuity programmes	✔	✔
13	Tie-ins	×	×
	Total Ticks	4	9

Legend: (✔) Indicates the presence of the programme.
(×) Indicates the obscene of the programme.

Notes: 1 Percentage score of checklist = $\frac{\text{No. of ticks}}{\text{No. of programmes}} \times 100$

2. Index for Vijaya dairy = 30.76 %
Index for Nandi dairy = 66.66 %

Source: Compiled from the records of Vijaya and Nandi dairy, Nandyal.

The percentage score of checklist is 30.76 per cent for Vijaya dairy and 66.66 per cent for Nandi dairy. From the above analysis it can concluded that level of sales promotion is higher in Nandi dairy than in Vijaya dairy.

EVENTS/EXPERIENCES PROGRAMS

Another component of promotion mix is events/experienced organized to popularize the dairy products. The individual programs are explained among in this section.

Table 7.3 shows events and experiences in Vijaya and Nandi dairies. Like the early table in this chapter, this table is also designed using checklist method. Out of nine events and experiences not a single event or experience was noticed in Vijaya dairy. In contrast out of nine events and experiences, five events and experiences were observed. Percentage score of checklist is zero for Vijaya dairy and 37.5 percent for Nandi dairy.

Table 7.3. Events/Experiences Programmes in Vijaya and Nandi Dairies

Sl. No.	Events/Experiences	Vijaya	Nandi
1	Sports	×	✔
2	Entertainment	×	×
3	Festivals	×	✔
4	Arts	×	✔
5	Causes	×	✔
6	Factory tours	×	×
7	Company museums	×	×
8	Street activities	×	✔
	Total ticks	...	5

Legend: (✔) indicates the presence of the programme.
(×) Indicates the obscene of the programme

Notes: 1 Percentage score of checklist $= \frac{\text{No. of ticks}}{\text{No. of programmes}} \times 100$

2. Index for Vijaya dairy = 0 %
Index for Nandi dairy = 37.5 %

Source: Compiled from the records of Vijaya and Nandi dairy, Nandyal.

Based on the above analysis it can be concluded that events and experiences were non-existence in promotion of dairy products in Vijaya dairy. Whereas half of the possible events and experiences were noticed in Nandi dairy. Thus, Nandi dairy has better promotion effort in terms of event/ experiences, compared to Vijaya dairy.

PUBLIC RELATIONS

Yet another element of promotion mix is public relations. It is non-paid form of promotion, in contrast to advertisement which is paid form of promotion. This is discussed in detailed in this section.

Table 7.4 presents public relations practiced in sample dairy units. The ten possible public relations events are identified and shown in column two of the table.

Table 7.4. Public Relations in Vijaya and Nandi Dairies

Sl. No.	Public relations	Vijaya	Nandi
1.	Press kits	×	✔
2.	Speeches	✔	×
3.	Seminars	✔	✔
4.	Annual reports	✔	✔
5.	Charitable donations	×	✔
6.	Publications	✔	×
7.	Community relations	×	✔
8.	Lobbying	✔	×
9.	Identity media	✔	✔
10.	Company magazine	×	×
	Total ticks	6	6

Legend: (✔) indicates the presence of the programme
(×) Indicates the obscene of the programme

Notes: 1. Percentage score of checklist = $\frac{\text{No. of ticks}}{\text{No. of programmes}} \times 100$

2. Index for Vijaya dairy = 60 %
Index for Nandi dairy = 60 %

Source: Compiled from the records of Vijaya and Nandi dairy,

Against each event either tick or cross mark is indicated, showing the presence or obscene of the event, there are six ticks for Vijaya dairy and six for Nandi dairy. When the check list is converted into percentage scores sixty is the per cent score for both the dairies. Going by preceding analysis it can be concluded that both the dairies have same level of public relations effort in promoting their dairy products.

Personal Selling

Personal selling is done by organizational employees who come in face-to-face contact with consumers. Variety of inducements is offered in personal selling to with over the potential consumers. Detailed presentation of these methods form part of this section Personal selling practices of the two sample dairy units are presented in Table 7.5. Out of possible

Table 7.5. Personal Selling Programmes in Vijaya and Nandi Dairies

Sl. No.	Personal selling	Vijaya	Nandi
1	Sales presentations	✔	✔
2	Sales meetings	✔	✔
3	Incentive programmes	×	✔
4	Samples	×	✔
5	Fairs and trade	✔	✔
6	Shows	✔	✔
	Total ticks	4	6

Legend: (✔) indicates the presence of the programme
(×) Indicates the obscene of the programme

Notes: 1. Percentage score of checklist = $\frac{\text{No. of ticks}}{\text{No. of programmes}} \times 100$

2. Index for Vijaya dairy = 66.66%
Index for Nandi dairy = 100%

Source: Compiled from the records of Vijaya and Nandi dairy,

six personal selling programmes, four programmes are adopted in Vijaya dairy and six in Nandi dairy. Converting check-list into percentage scores, Vijaya dairy has 66.66 and

Nandi dairy has hundred per cent score. Based on the above analysis it can be inferred that Nandi dairy has full-fledged personal selling programmes compared to that of Vijaya dairy.

DIRECT MARKETING

Directing marketing Practices involve the no intermediaries between the product selling dairy and consumers. Table: 7.6 direct marketing practices adopted by the two sample dairy units are exhibited in table 7.6. Out of eight possible direct marketing programmes, only one programme is adopted in each of these sample dairy units. Conversion of check-list yields 12.5 per cent score for both the dairies. Based on the low magnitude of the score, it can be concluded that direct marketing practices are at low level in both the dairies.

Table 7.6. Direct Marketing in Vijaya and Nandi Dairies

Sl. No.	Direct marketing	Vijaya	Nandi
1	Catalogs	✔	✔
2	Mailings	×	×
3	Telemarketing	×	×
4	Electronic shopping	×	×
5	TV shopping	×	×
6	Fax mail	×	×
7	E-mail	×	×
8	Voice mail	×	×
	Total ticks	1	1

Legend: (✔) indicates the presence of the programme

(×) Indicates the obscene of the programme

Notes: 1. Percentage score of checklist = $\frac{\text{No. of ticks}}{\text{No. of programmes}} \times 100$

2. Index for Vijaya dairy = 12.5%

Index for Nandi dairy = 12.5%

Source: Compiled from the records of Vijaya and Nandi dairy,

RANK ORDER OF COMMUNICATION MIX

Priorities accorded to communication mix by sample dairies are presented in this section.

Table 7.7. Rank Order of Communication mix in Vijaya and Nandi Dairy

Sl. No.	Communication mix	Rank order in	
		Vijaya Dairy	Nandi Dairy
1	Advertising	1	2
2	Sales Promotion	6	3
3	Events/Experiences	4	5
4	Public relations	3	4
5	Personal Selling	2	1
6	Direct marketing	5	6

Source: Primary data

Communication mix elements are rank-ordered for the two sample dairy units and presented Table 7.7. Direct marketing is the number one element of communication mix of Vijaya dairy units followed by public relations and sales promotions. Out of six elements of communication-mix events or experiences rank least in Vijaya dairy. On contrary sales promotion is assigned rank one in Nandi dairy followed by direct marking, advertising and events or experiences. Like in Vijaya dairy events or experiences are ranked least in Nandi dairy as an element of communication.

Communication Budgets and Sales

Huge investments are made by dairy on consumer communication, adopting different methods for deciding the volume of the budgets.

Table 7.8 are drawn to present methods used for deciding total communication budget in the two sample dairies. Four such possible methods identified are shown in column two namely affordable methods, percentage method, and

Table 7.8. Methods Used for Determining total Communication Budgets in Vijaya and Nandi Dairies

Sl. No.	Method used for deciding total communication budgets	Vijaya dairy	Nandi dairy
1	Affordable method	✔	✔
2	Percentage method	×	×
3	Competitive method	✔	✔
4	Objective and task method	✔	×
	Total ticks	3	2

Legend: (✔) indicates the presence of the programme
(×) Indicates the obscene of the programme

Notes: 1. Percentage score of checklist = $\frac{\text{No. of ticks}}{\text{No. of programmes}} \times 100$

2. Index for Vijaya dairy = 75%
Index for Nandi dairy = 50%

Source: Compiled from the records of Vijaya and Nandi dairy,

competitive method, objective and task method. Out of the four methods, three methods are verified against Vijaya dairy and two against Nandi dairy. With the conversion of check-list into percentage scores 75 per cent is the score for Vijaya dairy and 50 per cent for Nandi dairy. Relatively Vijaya dairy used more methods for deciding total communication budget as compared to Nandi dairy.

Table 7.9 presents communication budget for 10 year period from 1998-2007 for sample dairy units. As can be seen from the table average annual communication budget is Rs 2.39 lakh in Nandi dairy which is marginally higher than the annual budget of Rs 2.18 lakh in Vijaya dairy. LGRs are of higher magnitude at 13.54 per cent and 14.78 per cent in Vijaya dairy and Nandi dairies respectively. Calculate t-values are 6.73 and 8.32 per cent in Vijaya dairy and Nandi Dairy respectively, which are, Significant at 1% level. Calculate 'r'-values are 0.9217 for Vijaya dairy and 0.1468 for Nandi dairy, which indicate this slope of regression lines.

Table 7.9. Yearly Total Communication Budget in Vijaya and Nandi Dairies

Sl. No.	Year	Budget (Rs. in lakhs)	
		Vijaya	Nandi
1	1998	1.20	1.00
2	1999	1.35	1.50
3	2000	1.50	1.50
4	2001	1.80	1.90
5	2002	1.50	2.00
6	2003	2.00	2.00
7	2004	2.00	2.50
8	2005	3.00	3.00
9	2006	3.50	4.00
10	2007	4.00	4.50
	Avg	2.18	2.39
	LGR	13.54	14.78
	/t/ value	6.73**	8.32**
	r. value	0.9217	0.9468

Note: **Significant at 1 per cent level.
Source: Compiled from the records of Vijaya and Nandi dairy.

Table 7.9 and 7.10 show the are shown annual sales and total communication budgets of Vijaya and Nandi dairy units for 10 year period. Reading the last rows of the table, it can be seen that annual average communication budget of Nandi dairy is higher at Rs. 2.39 lakh than that of Vijaya dairy (Rs 2.18 lakh). In contrast annual average sales of Vijaya dairy are higher at Rs. 23.684 crore than that of Nandi dairy (Rs. 21.50 crore). LGRs of both annual sales total communication budgets are of higher magnitude in Nandi dairy than Vijaya dairy, which are statistically significant at one percent level, Linear regression results of annual sales on total communication shows that regression coefficients are Rs. 910.26 lakh and Rs. 769.59 lakh for Vijaya and Nandi dairies respectively, which mean that Rs. 1 lakh annual spending communication brings Rs. 910.26 lakh sales in Vijaya dairy and Rs. 769.59 lakh in Nandi dairy 'r'^{2}-values of 0.71 and 0.88 for Vijaya and Nandi dairies mean 71 and 88 percent of annual sales variation of respectively explained by annual spending on communication.

Table: 7.10. Yearly Total Communication Budgets and Sales in Vijaya and Nandi Dairies

(Rs. in lakhs)

S. No.	Year	Vijaya Annual total communication budget	Nandi Annual sales	Annual total communication budget	Annual sales
1	1998	1.20	1371.22	1.00	894.95
2	1999	1.35	1358.95	1.50	898.35
3	2000	1.50	1424.03	1.50	1347.45
4	2001	1.80	1432.09	1.90	1850.32
5	2002	1.50	3153.72	2.00	1950.61
6	2003	2.00	1845.35	2.00	2332.46
7	2004	2.00	2395.83	2.50	2616.32
8	2005	3.00	2781.35	3.00	2878.87
9	2006	3.50	3873.80	4.00	3083.49
10	2007	4.00	4045.60	4.50	3627.14
	Avg.	2.18	2368.00	2.39	2147.99
	LGR	13.54	12.80	14.78	14.17
	/t/ value	6.73 **	5.10 **	8.32 **	23.86 **
	r. value	0.9217	0.8745	0.9468	0.9930

Note: **Significant at 1 per cent level.

Source: Compiled from the records of Vijaya and Nandi dairy

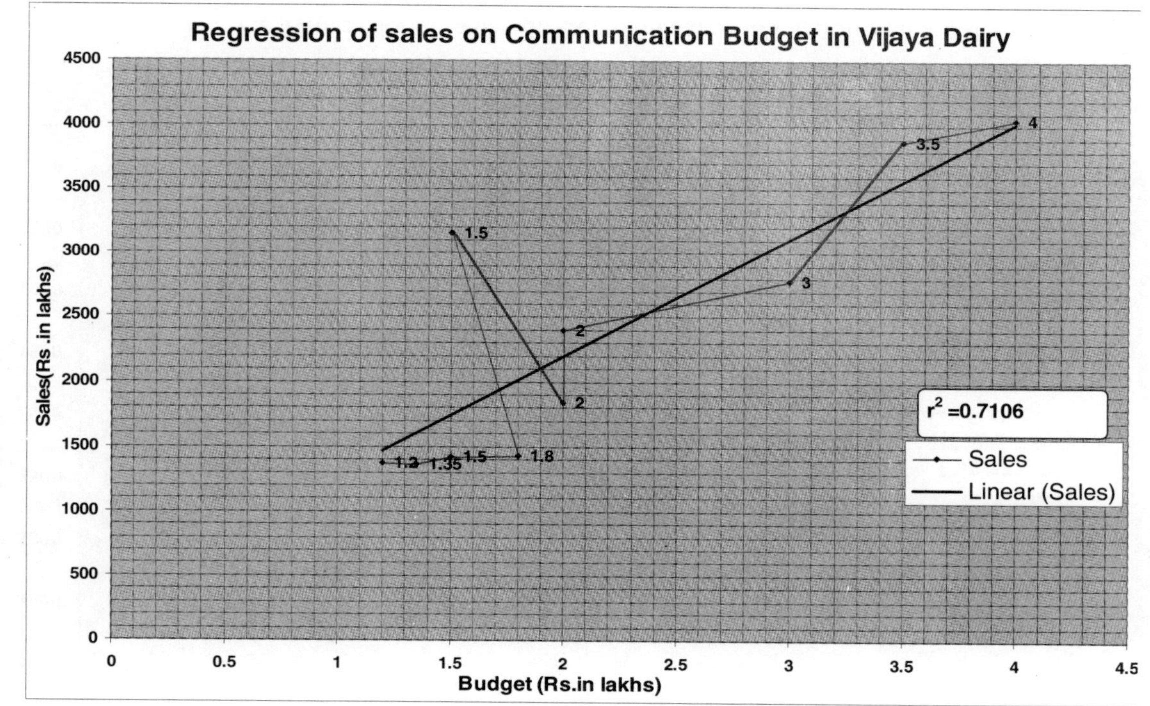

Fig. 7.1. Regression of sales on Communication Budget in Vijaya Dairy

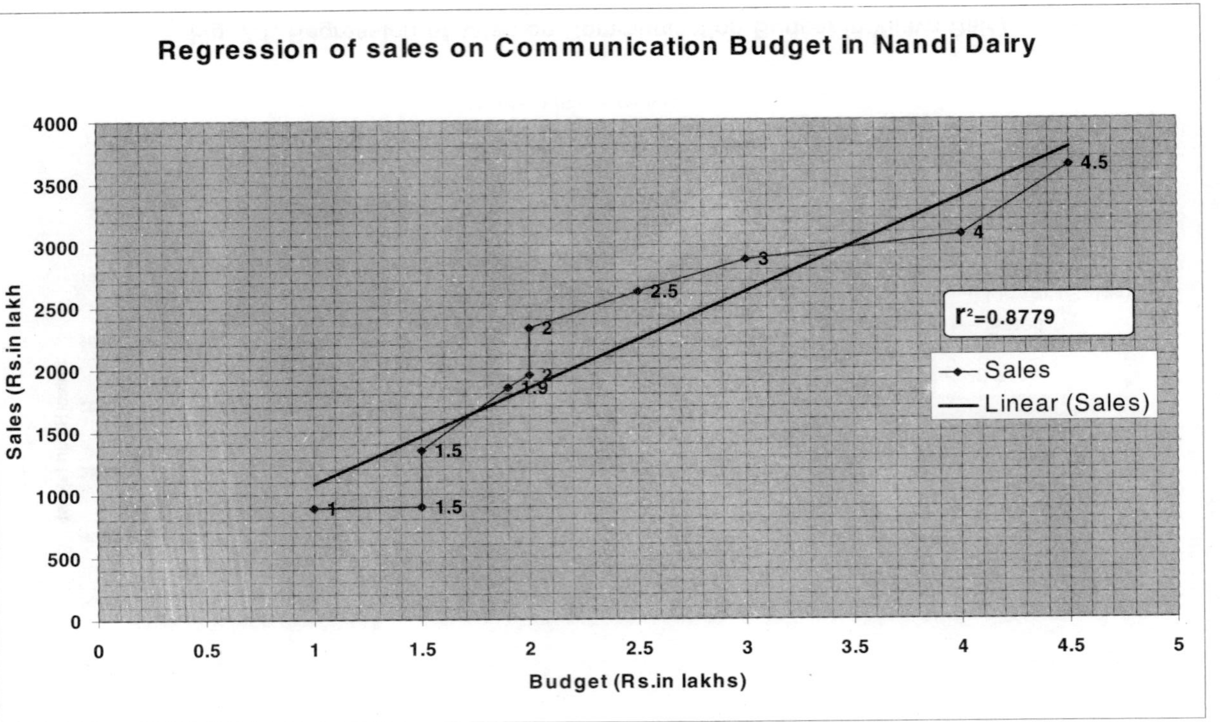

Fig. 7.2. Regression of sales on Communication Budget in Nandi Dairy

Advertising Budgets and Sales

Advertising is sales paid form of communication, targeting consumers and using different media in Table 7.11 are shown the advertising budgets of Vijaya and Nandi dairies for 10 year period from 1998-2007. Average annual budgets are Rs. 2.55 lakh in Vijaya dairy and Rs. 2.51 lakh in Nandi dairy, which are of almost of same magnitude. The LGR of advertising budget is 12.66 per cent in Vijaya dairy which is one per cent higher than LGR of 11.61 per cent of Nandi dairy. Calculated t-values are statistically significant at 1 per cent level. Calculated 'r'-values of Vijaya and Nandi dairy are 0.9887 and 0.9616 respectively.

Table: 7.11. Yearly Total Advertising Budgets in Vijaya and Nandi Dairies

(Rs. in lakhs)

Sl. No.	Year	Budget	
		Vijaya dairy	Nandi dairy
1	1998	1.20	1.50
2	1999	1.50	1.50
3	2000	1.50	1.80
4	2001	1.90	2.00
5	2002	2.50	2.20
6	2003	2.70	2.50
7	2004	3.20	2.60
8	2005	3.50	3.00
9	2006	3.50	4.00
10	2007	4.00	4.00
	Avg	2.55	2.51
	LGR	12.66	11.61
	/t/ value	18.73 **	9.85 **
	r. value	0.9887	0.9616

Note: **Significant at 1 per cent level.
Source: Compiled from the records of Vijaya and Nandi dairies, Nandyal.

Table 7.12 exhibits annual sales and expenditure on advertising in Vijaya and Nandi dairies for a decade-period. Annual advertising expenses are marginally higher in Vijaya dairy at Rs. 2.55 lakh than in Nandi dairy at Rs. 2.51 lakh. Annual LGR is higher in Nandi dairy (14.17%) than in Vijaya dairy (21.15%), which are highly significant at one per cent level. Linear regression of sales run on annual advertising expenses reveal that 'r' values are 926.84 and 268.91 for Vijaya and Nandi dairies respectively, which mean that Rs. 1 lakh annual spending on advertising brings Rs. 9.26 crore

Table 7.12. Yearly Total Advertising Budgets and Sales in Vijaya and Nandi Dairies

(Rs. in lakhs)

Sl. No.	Year	Annual total advertising budget	Annual sales	Annual total advertising budget	Annual sales
1	1998	1.20	1371.22	1.50	894.95
2	1999	1.50	1358.95	1.50	898.35
3	2000	1.50	1424.03	1.80	1347.45
4	2001	1.90	1432.09	2.00	1850.32
5	2002	2.50	3153.72	2.20	1950.61
6	2003	2.70	1845.35	2.50	2332.46
7	2004	3.20	2395.83	2.60	2616.32
8	2005	3.50	2781.35	3.00	2878.87
9	2006	3.50	3873.80	4.00	3083.49
10	2007	4.00	4045.60	4.00	3627.14
	Avg	2.55	2368.00	2.51	2147.99
	LGR	12.66	12.80	11.61	14.17
	/t/ value	18.73 **	5.10 **	9.85 **	23.86 **
	r. value	0.9887	0.8745	0.9616	0.9930

Note: **Significant at 1 per cent level.

Source: Compiled from the records of Vijaya and Nandi dairy

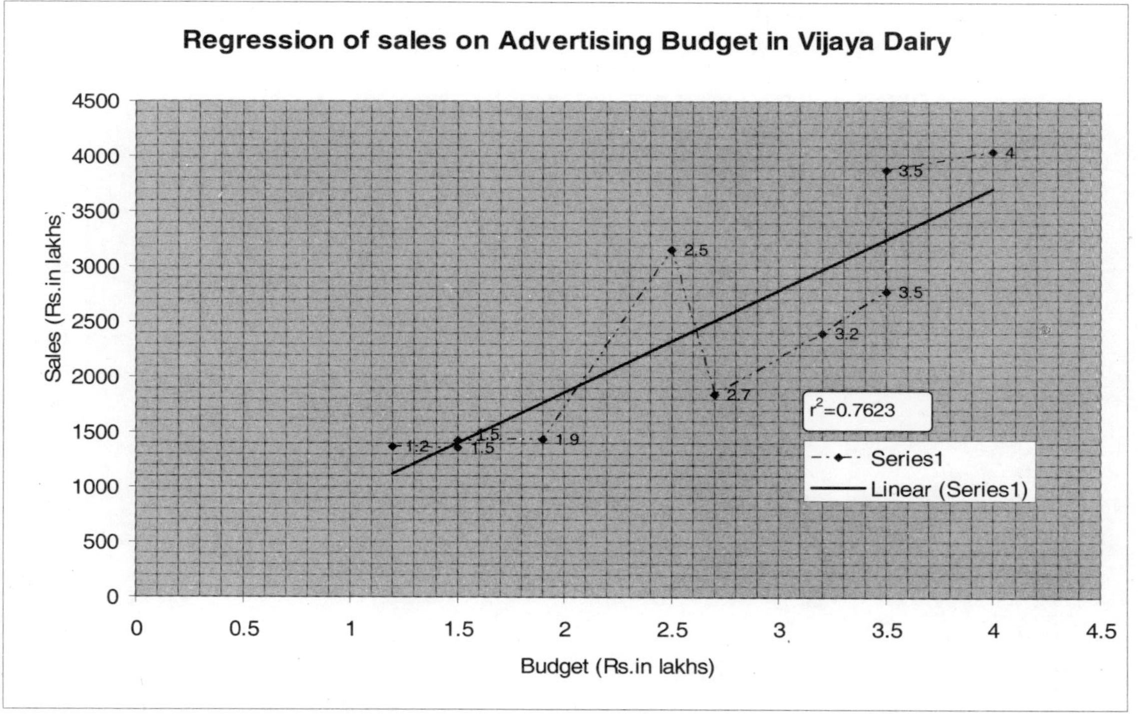

Fig. 7.3. Regression of sales on Advertising Budget in Vijaya Dairy

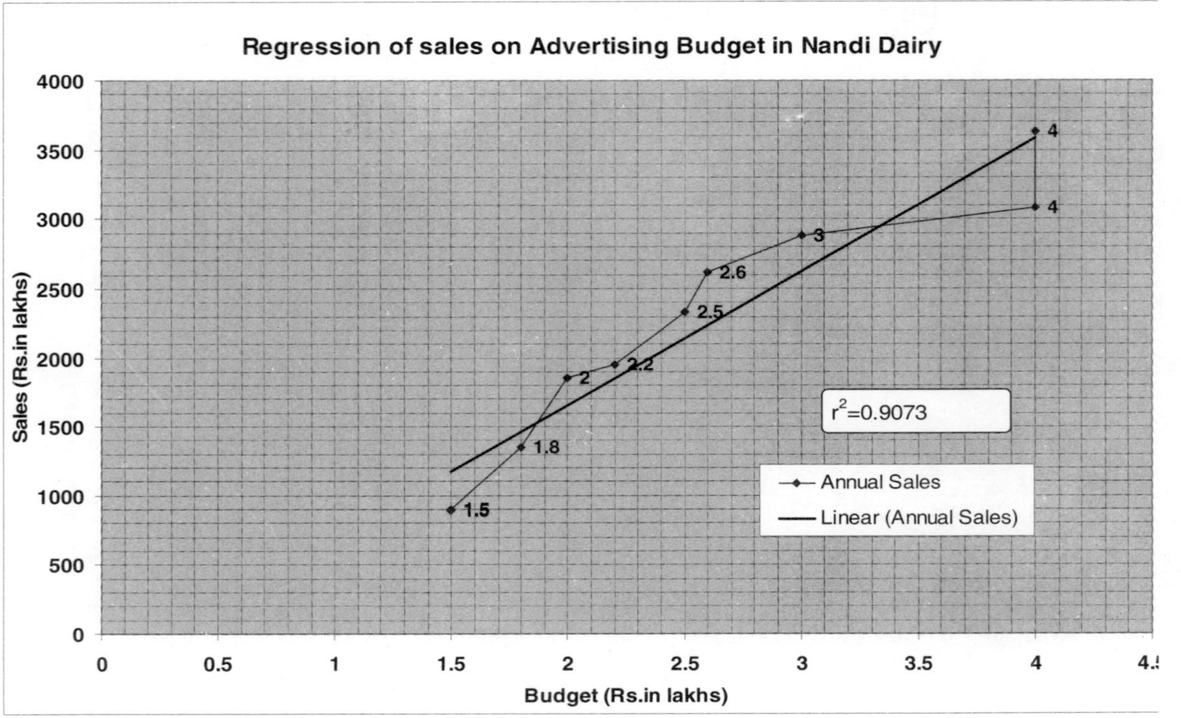

Fig. 7.4. Regression of sales on Communication Budget in Nandi Dairy

annual sales in Vijaya dairy and Rs. 2.68 crore in Nandi dairy. "r'2-values are of higher magnitude in both the dairies-0.76 in Vijaya dairy and 0.97 in Nandi dairy, which mean that annual sales variations to the extent 76 per cent and 97 per cent in Vijaya and Nandi dairies respectively are explored by annual spending on advertisement.

Sales Promotion

Variety of inducements is offered to the consumers in sales promotion. Table 7.13 shows the sales promotion tools

Table 7.13. Dairy Product Sales Promotion Tools in Vijaya Dairy

Sl. No.	Sales promotion	Products	No. of products
1	Samples	Flavoured milk	1
2	Coupons	Ghee	1
3	Frequently Programmes	Double Toned Milk Toned Milk Whole Milk	3
4	Prizes	Double Toned Milk Toned Milk Whole Milk	3
5	Free tails	Butter milk	1
6	Point of Purchasesd/ Displays and demonstrations	(1) Milk Products (a) Skim Milk (b) Double Toned Milk (c) Toned Milk (d) Standardized Milk (e) whole Milk (2) Basic Product (a) Pannir (3) By Products (a) Butter (b) Butter Milk (c) Ghee (d) Skimmed milk powder (e) Doodh peda (f) Sterilized flavoured milk	12

Source: Compiled from the records of Vijaya dairy.

adopted by Vijaya dairy unit during 2006-07 are exhibited in table: 7.13. Altogether six sales promotion tools are adopted such as sample coupons, frequently programmes, prizes, free tails and point of purchase display and demonstration. In the last column of the table is shown the number of product which was covered under each of the sales promotion programmes. Twelve products are covered under point of purchases display and demonstration; three products each were under covered prizes and frequently programmes, one product each under sample coupons and free tails programmes. Based on above analysis it can be concluded that point of purchase display and demonstrations programmes is the all product covering programme followed by prizes.

Table 7.14 is designed on the lines of preceding table to demonstrate sales promotion programmes implemented in Nandi dairy. There are six programmes similar to Vijaya dairy implemented in Nandi dairy. The individual products covered under the each programme and their numbers are shown in column three and four of the table, respectively. Like in Vijaya dairy point of purchases display and demonstration is all product covering programme, of under which thirteen individual products are covered next come frequently programmes and prizes under which three products each are covered Sample coupons and free tails covered product each. In comparison two dairy units are similar in their sales promotion programmes.

Media-wise Advertising Budgets

Relative importance of media in which advertisements are inserted, is reflected in size of advertisement of budgets in the reference period.

Table 7.15 exhibits annual advertising budget in different media for ten-year period from 1998 to 2007. The different media used for advertising are news papers, television, magazines, yellow pages/newsletters, broachers/pamphlets. Focusing on the column 10 of table, it can be seen that budget

Table 7.14. Dairy Product Sales Promotion Tools in Nandi Dairy

Sl. No.	Sales promotion	Products	No. of products
1	Samples	Butter milk	1
2	Coupons	Ghee	1
3	Frequently Programmes	Double Toned Milk Toned Milk Gold Milk	3
4	Prizes	Double Toned Milk Toned Milk Gold Milk	3
5	Free tails	Butter milk	1
6	Point of Purchasesd/ Displays and demonstrations	(1) Milk Products (a) Skim Milk (b) Double Toned Milk (c) Toned Milk (d) Standardized Milk (e) Gold Milk (2) By-products (a) Ghee (b) Butter-milk (3) Sweet Product (a) Khawvva (b) Kalakan (c) Doodhpeda (d) Milk Mysur Pack (e) Palakova (f) Makhan	13

Source: Compiled from the records of Nandi dairy

utilization as percentage of budget allocation was highest at 116 per cent in 1999 and longest at 92.88 per cent in 2003. In four out of ten years the budget utilization exceeded budget allocation. Concentrating on average row at bottom of the table it can be observed the annual average budget utilization for ten year period is highest on news papers at Rs1.17 lakh followed by Rs. 0.98 lakh on broachers or pamphlets Rs. 0.97 lakh in yellow pages/news letters Rs. 0.93 lakh in television and Rs. 0.28 lakh on magazines. LGRs of

Table 7.15. Annual Advertising Budgets of Different Media in Vijaya Dairy

(Rs. in lakhs)

Sl. No.	Years	Annual media advertising budget	Budget utilization					Total budget utilization (4 to 8)	Column 9 as %ge of column 3
			Newspapers	Television	Magazines	Yellow Pages/ Newsletters	Broachers/ Pamphlets		
1	2	3	4	5	6	7	8	9	10
1	1998	3.00	0.70 (23.72)	0.30 (10.16)	0.20 (6.77)	0.70 (23.72)	1.05 (35.59)	2.95 (100)	98.33
2	1999	2.50	0.75 (25.86)	0.35 (12.06)	0.20 (6.89)	0.80 (27.58)	0.80 (27.58)	2.90 (100)	116.00
3	2000	3.00	0.78 (24.60)	0.32 (10.09)	0.22 (6.94)	0.85 (26.81)	1.00 (31.54)	3.17 (100)	105.66
4	2001	3.00	0.85 (29.51)	0.38 (9.72)	0.25 (8.68)	0.90 (31.25)	1.00 (34.72)	2.88 (100)	96.00
5	2002	3.00	1.00 (32.67)	0.40 (13.07)	0.26 (8.49)	0.90 (29.41)	0.50 (16.33)	3.06 (100)	102.00
6	2003	3.50	1.10 (31.97)	0.42 (12.20)	0.22 (6.39)	0.90 (26.16)	0.80 (23.25)	3.44 (100)	98.28
7	2004	3.50	1.30 (33.76)	0.46 (11.94)	0.24 (6.23)	1.00 (25.97)	0.85 (22.07)	3.85 (100)	110.00

1	2	3	4	5	6	7	8	9	10
8	2005	4.50	1.45 (34.68)	0.42 (10.04)	0.26 (6.22)	1.10 (26.31)	0.95 (22.72)	4.18 (100)	92.88
9	2006	5.00	1.80 (36.73)	0.40 (8.16)	0.40 (8.16)	1.20 (24.48)	1.10 (22.44)	4.90 (100)	98.00
10	2007	6.00	1.95 (32.82)	0.44 (7.40)	0.50 (8.41)	1.30 (21.88)	1.75 (29.46)	5.94 (100)	99.00
	Avg	3.70	1.17	0.39	0.28	0.97	0.98	3.73	
	LGR	8.84	12.14	3.69	9.32	6.12	4.94	8.38	
	/t/ Value	5.59 **	10.22 **	4.39 **	3.70 **	10.19 **	1.46 NS	5.44 **	
	r. Value	0.89	0.96	0.84	0.79	0.96	0.45	0.89	

Note: ** Significant at 1 % level, NS: Not significant at 5 % level
Source: Compiled from the records of Vijaya dairy, Nandyal.

Table: 7.16. Annual Advertising Budgets of Different Media in Nandi Dairy

(Rs. in lakhs)

Sl. No.	Years	Annual media advertising budget	Budget utilization					Total budget utilization (4 to 8)	Column 9 as %ge of column 3
			Newspapers	Television	Magazines	Yellow Pages/ Newsletters	Broachers/ Pamphlets		
1	2	3	4	5	6	7	8	9	10
1	1998	3.00	0.70 (23.72)	0.50 (16.94)	0.40 (13.55)	0.50 (16.94)	0.85 (28.81)	2.95 (100)	98.33
2	1999	3.00	1.00 (31.25)	0.30 (9.37)	0.45 (14.60)	0.55 (17.18)	0.90 (28.12)	3.20 (100)	106.66
3	2000	4.00	1.10 (29.72)	0.40 (10.81)	0.50 (13.51)	0.70 (18.91)	1.10 (29.72)	3.70 (100)	92.50
4	2001	4.00	1.20 (29.26)	0.60 (16.69)	0.40 (9.75)	0.75 (18.29)	1.15 (28.04)	4.10 (100)	102.50
5	2002	5.00	1.10 (23.75)	0.45 (9.71)	0.58 (12.54)	0.90 (19.43)	1.60 (34.55)	4.63 (100)	92.60
6	2003	5.00	1.30 (24.90)	0.62 (11.87)	0.60 (11.49)	1.05 (20.11)	1.65 (31.60)	5.22 (100)	104.40
7	2004	5.00	1.40 (26.87)	0.66 (12.66)	0.65 (12.47)	1.10 (12.47)	1.40 (26.87)	5.21 (100)	104.20

1	2	3	4	5	6	7	8	9	10
8	2005	5.50	1.20 (20.97)	0.68 (11.87)	0.64 (11.08)	1.40 (12.47)	1.80 (31.46)	5.72 (100)	104.00
9	2006	6.00	1.50 (24.87)	0.55 (9.12)	0.68 (11.27)	1.45 (11.27)	1.85 (30.67)	6.03 (100)	100.50
10	2007	6.50	1.60 (24.80)	0.60 (9.30)	0.70 (10.85)	1.50 (10.85)	2.05 (31.78)	6.45 (100)	99.23
	Avg	4.70	1.21	0.54	1.19	0.99	1.43	4.72	
	LGR	8.12	6.46	4.97	31.82	12.24	9.18	8.38	
	/t/ Value	12.96 **	6.20 **	2.48 *	1.91 NS	18.99 **	8.84 **	27.00 **	
	r. Value	098	0.90	0.66	0.56	0.99	0.95	0.99	

Note: ** Significant at 1 % level, NS: Not significant at 5 % level
Source: Compiled from the records of Nandi dairy, Nandyal.

budget spending on different media are significant at except 1per cent level broachers and pamphlets. Thus, percentage of utilization as percentage of budget of allocation budget as percentage has been very high in Vijaya dairy. Expenditure on newspaper was in highest and on magazine it is least.

Table 7.16 is drawn to present advertising budget on different media in Nandi dairy from 1998 to 2007. The budget utilization as percentage of allocation on advertising was highest at 106 per cent in 1999 and lowest at 92 per cent in 2002. Out of 10year period under reference the budget utilization more has been 100 per cent in 6 years. Coming to average row at bottom of the table it could be seen the annual average expenses on advertising in broachers/pamphlets was highest at Rs 1.43 lakh and lowest at Rs. 0.54 lakh on television. LGRs are significant for total budget allocation news papers, media, yellow pages/newsletters/pamphlets and budget utilization at one per cent level. LGRs are significant at 5 per cent level with regard to budget spending on television. In contrast LGR is not significant with respective to spending on advertisement in magazines. From the above analysis it can be concluded that budget utilization has been relatively higher in Nandi dairy unit as compared to Vijaya dairy. Unlike Vijaya dairy advertisement expenses are highest on brooches/ pamphlets lowest on Television.

CONCLUSION

Advertising programmes are of equal importance in both the dairies, going by the number of programmes undertaken. The budget of sales promotion programmes taken up in the sample dairies, there are marked variations between them. Similar in the case events/experiences programmes, public relations are of high order in both the dairies. Degree of personal selling is more in both the dairies as compared to direct marketing. LGRs of communication advertising budgets as well as sales are of higher magnitudes which are significant too point of purchase displays/demonstrations are by far the most adopted sales promotion technique in both the dairies. Advertisements are inserted in wide range of media in both the dairies.

8 CHAPTER

SALES AND DISTRIBUTION PRACTICES OF DAIRIES

INTRODUCTION

Place, which is the last 'P' of 4 P's-marketing conceptual frame work is dealt in this chapter. It mainly focuses on channels of distribution, push-pull strategies, sales through retail outlets and dealers, sales to institutions and direct sales. Each of these aspects is analysed in the sections to follow.

Marketing Channel Systems and Strategies

Channel players and their functions ultimately decide the distribution effectiveness and efficiency of the dairy. This section exclusively dealers with these aspects. Channel levels and channel players are shown in Table 8.1. It can be seen that the diary and consumer and milk product are part every channel. The number of intermediary levels is the basis for the designating the length of the channel. Zero level channel or direct selling to the diary products consumer. In one level channel stockists are intermediaries linking the diary and consumer. In two level channel there are present to intermediaries mainly dealer/agent and stockists in between the diary and the consumers.

In managing the inter media over involved in marketing the dairy products, Vijaya and Nandi dairies are involved in decisions as to how much is devoted to pull verses push marketing strategies. To make these concepts clear, this can be made explicit and elaborate. Push strategy involves the diaries using their sales force and trade promotions money so as to induce to their intermediaries to carry, promote and sell the diary products to end users. Push strategies are adopted when the brand loyalty is low, brand choice is made in the store and product is an impulsive purchase product benefits are well understood log the consumers. In contrast, in diary marketing push strategy involves the dairies using advertising and promotion persuades diary product consumers to ask intermediaries for the products, this inducing the intermediaries to order them. this strategies are found appropriate when there is high brand loyalty and high customer involvement in this categories, when customers perceive differences between brands, when customers choose the brand to going to shop.

Table 8.1. Marketing Channel Systems in Vijaya and Nandi Dairies

Sl. No.	Channel level	Channel descriptions
1.	Vijaya dairy	
	(a) O-Level	Dairy → Consumer
	(b) 1-Level	Dairy → Dairy Stockists → Consumer
	(c) 2-Level	Dairy→ Dealer/Agent→ Stockists → Consumer
2.	Nandi dairy	
	(a) O-Level	Dairy→Consumer
	(b) 1-Level	Dairy→Dairy Stockists→Consumer
	(c) 2-Level	Dairy→ Dealer/Agent→Stockists→Consumer

Source: Primary Data.

Thus there are three levels of channel in operation in both Vijaya and Nandi units. The numbers of channel levels are limited in these diaries, resulting in no problems as to consumers' information and control over channel members.

From what can be seen in Table 8.2 it can be understood that both the dairies are found practising push and pull strategies in diary products marketing. With the diaries have common pull strategies falling under each of push and pull strategies? Players in push strategies are company outlets, dealers, direct-selling and marketing agents. Under pull strategies come and advertisements, publicity and public relations. From the preceding analysis it can be concluded that the diaries execute push as well as pull strategies.

Table 8.2. Push-Pull Strategies in Vijaya and Nandi Dairies

Sl. No.	Name of the dairy	Push strategies /players	Pull strategies
1.	Vijaya	(a) Dairy sales Outlets (b) Dealers (c) Direct Selling (d) Marketing agents	(a) Advertisements (b) Publicity (c) Public relations
2.	Nandi	(a) Dairy sales Outlets (b) Dealers (c) Direct Selling (d) Marketing agents	(a) Advertisements (b) Publicity (c) Public relations

Source: Primary Data

Sales through Dairy Owned Retail Outlets

One obstinct feature of dairy marketing is sales made through dairy-owned retail outlets. This is so because milk and milk product meant for level consumption are sold through the retail outlets. As can be seen Table 8.3 that Vijaya dairy sells milk through direct and indirect channels as explained earlier. The company has established its own retail outlets to distribute milk in its market. There were three outlets in each of 1998 and 1999. They had risen to four in the following year and five in 2001 which continuous to be stale till 2007. It continued the sales through five outlets only. In other words, there was no addition to out lets from 2001 to 2007. The retail outlets sold 17.23 lakh litres in 1998, 2.31 per cent of total sales. By 2007 the sale of milk increased to 39.00 lakh litre with relative ups and downs over the period. The year 2000 recorded the lowest sales at 14.78 lakh litres,

sales through retail outlets constituted 20.21 per cent total sales in 2007. The proportion of retail outlets in the aggregates sales varied between 15.67 per cent in 1998 and 25.88 per cent in 2005. From the preceding analysis it can be concluded that proportion of retail outlets in the total sales even at the pack never crossed 26 per cent growing trend in absolute sales, is due to the fact that the growth in sales through other than retail outlet has been growing much fasters.

Table 8.3. Dairy Retail Outlets and Growth of Sales in Vijaya Dairy

Sl. No.	Year	No. of dairy sales outlets	Sales through outlets (lakh litres)	Total sales of a dairy	%age of col. (4) to col. (5)
1	1998	3	17.32	77.62	22.31
2	1999	3	16.87	73.82	22.85
3	2000	4	14.78	75.13	19.67
4	2001	5	15.52	76.37	20.32
5	2002	5	27.58	168.08	16.40
6	2003	5	17.11	94.41	18.12
7	2004	5	18.40	117.351	5.67
8	2005	5	35.70	137.932	5.88
9	2006	5	48.76	193.06	25.25
10	2007	5	39.00	192.95	20.21

Source: Primary Data.

As can be noticed in Table 8.4 in 1998 the Nandi dairy sold 3.05 lakh litres of the milk through three retail outlets where share in the total sales is 5.57 per cent in the aggregate sale of 54.75 lakh litres milk in 1998, in the tyrannical year aggregate sales reached 156.95 litres, of which sales through dairy outlets amounted to 16.30 lakh litres constituting 10.38 per cent of total sales. The retail outlets were four in 2000 where numbered rose to five in the following years and continued as such till 2007. The retail outlets sales through experienced marked inverse in the past three years under reference. It can be concluded that there is a significant growth in the volume of milk sales despite in significant growth in the number of outlets. In relative terms, their sales improved remarkable from 5.57 per cent in 1998 to 10.38 per cent in 2007. The performance of retail outlets of Nandi dairy cannot be favourably compared with that of Vijaya dairy. This may be on account of market differences between the two dairies, the Vijaya dairy wider coverage and the Nandi warrens geographical coverage.

Table 8.4. Dairy Retail Outlets and Growth of Sales in Nandi Dairy

Sl. No.	Year	No. of dairy sales outlets	Sales through outlets (lakh litres)	Total sales of a dairy	% age of col. (4) to col. (5)
1	1998	3	3.05	54.75	5.57
2	1999	3	3.34	54.70	6.10
3	2000	4	3.10	73.00	4.24
4	2001	5	6.35	91.25	6.95
5	2002	5	2.60	94.90	2.73
6	2003	5	5.00	109.50	4.56
7	2004	5	4.00	116.80	3.42
8	2005	5	13.50	131.40	10.27
9	2006	5	13.80	138.70	9.94
10	2007	5	16.30	156.95	10.38

Source: Primary Data.

Distribution through Dealer Networks

Alternative channel of distribution of milk and milk production is dealer/stockiest who are dominant players. As can be seen from Table 8.5 that the Vijaya dairy used the services of 255 dealers/agents in 1998 distributing milk in the market. Whose number for increased to 410 in 2007 with relative fluctuating in the meantime. In the 1998, initial years, dealers/agents sold 40.15 lakh litres while in the terminal year, 2007, they sold 110.30 lakh litres. There are ups and downs in the sales in the reference period. Their share in the aggregate sales was 51.72 per cent in 1998 *vis-à-vis* 57.16 per cent in 2007. It can be summed up that has been that there growth in the number of dealers/agents and the volume of sales as well. Further more than 50 per cent of sales are effected by dealers/agents except 2006 (41.23 %) in all the years except in.

Table 8.5. Growth of Dairy Dealers and Sales in Vijaya Dairy: From 1998-2007

Sl. No.	Year	No. of dairy sales outlets	Sales through outlets (lakh litres)	Total sales of a dairy	% age of col. (4) to col. (5)
1	1998	255	40.15	77.62	51.72
2	1999	240	38.45	73.82	52.08
3	2000	245	39.05	75.13	51.97
4	2001	260	41.25	76.37	54.01
5	2002	325	98.10	168.08	58.36
6	2003	280	55.15	94.41	58.41
7	2004	310	65.45	117.35	55.77
8	2005	350	82.08	137.93	59.05
9	2006	380	95.60	193.06	49.23
10	2007	410	110.30	192.95	57.16

Source: Primary Data.

Presenting Table 8.6 it can be noticed that the Nandi dairy 200 dealers/agents involved in the channel of distribution of milk has whose number increased to 310 in 2007,the tyrannical year the number has been fluctuations over the period. These dealers/agents sold 41.50 lakh litres in 1998 as compared to while 112.55 lakh litres in 2007. There has been progressive increase in the quantum of sales expect a marginal decline in 1999. In the total sales, the dealers/ agents accounts for 75.79 per cent in 1998 of total sales as compared in 2007. While 1.71 per cent with ups and downs can be observed in the period. The year 2006 recorded the lowest share of 65.10 per cent of total sales. It may be inferred that the performance of dealers/agents in terms of their percentage share total sales is higher in Nandi dairy than in Vijaya dairy. This is evident from the fact that per dealer sales in 1998 was 20.750 litres and 15.745 in Nandi dairy and Vijaya dairy respectively. Similarly in the former and later dairies the dealers/agents average sales were 36.306 litres and 26.902 litres respectively.

Table 8.6. Growth of Dairy Dealers and Sales in Nandi Dairy

Sl. No.	Year	No. of dairy sales outlets	Sales through outlets (lakh litres)	Total sales of a dairy	%age of col. (4) to col. (5)
1	1998	200	41.50	54.75	75.79
2	1999	220	40.25	54.70	73.58
3	2000	230	51.40	73.00	70.41
4	2001	260	60.80	91.25	66.63
5	2002	240	70.20	94.90	73.97
6	2003	242	74.40	109.50	67.94
7	2004	246	78.20	116.80	66.95
8	2005	260	86.25	131.40	65.63
9	2006	272	90.30	138.70	65.10
10	2007	310	112.55	156.95	71.71

Source: Primary Data.

Instructional *vis-à-vis* total Sales

Milk is sold by sample dairies both to institutional and individual consumers. Relative proportion of institutional sales in total sales is discussed in this section.

Table 8.7 shows details of direct sales of Vijaya Dairy to institutional buyers during 1998-2007. This dairy he sold milk in bulk quantities to 25 institutions in 1998 as against 135 institutions in 2007. Gradual growth is instilled the number of institutional buyers except in years 2000, 2003 and 2005 which registered marginal declines over the respective immediately preceding years. The Vijay dairy to these institutions supplied 20.15 lakh letters in 1998 as compared to 43.65 lakh letters in 2007. In the mean time there are ups and downs the year 2006 recording the high quantum of sales of 48.70 percentages and the year 1998 registering the longest sales the least sales 18.70 lakh litres.

Table 8.7. Sales to Institutions *vis-à-vis* Total Sales in Vijaya Dairy

Sl. No.	Year	No. of institutions milk sold points	Sales to institutions (lakh litres)	Total sales (in lakh litres)	col (4) as % age of col (5)
1	1998	25	20.15	77.62	25.95
2	1999	30	18.50	73.82	25.06
3	2000	27	21.30	75.13	28.35
4	2001	38	20.60	76.37	26.97
5	2002	50	42.40	168.08	25.22
6	2003	45	22.15	94.41	23.46
7	2004	70	33.40	117.35	28.46
8	2005	62	20.15	137.93	14.60
9	2006	120	48.70	193.06	25.22
10	2007	135	43.65	192.95	22.62

Source: Primary Data.

These institutions sales in the total formed 25.95 per cent in 1998 *vis-a-vis* 22.62 per cent in 2007. In the mean while the fluctuations are noticeable. For instance 2004 showed the maximum share of 28.46 per cent whereas the least of 14.60 per cent was recorded in the following year. It can be summed up that despite an increase in the number of intuitional buyers and consequent increase in the quantum of sales, the proportion of direct sales in the total sales to institutions has declined in the recent past as compared the early period.

Table 8.8 shows the details of direct sales of Nandi Dairy institutions during 1998- 2007. Yearly direct sales mean milk sale of milk to institution like hospitals, schools, industrial units without involvement of middlemen in the channel of distribution between milk dairy and users. A look at the table reveals direct sales were effected to 25 institutions 1998 as against 110 in 2007.

Table 8.8. Sales to Institutions vis-à-vis Total Sales in Nandi Dairy

Sl. No.	Year	No. of institutions milk sold points	Sales to institutions (lakh litres)	Total sales (in lakh litres)	col (4) as % age of col (5)
1	1998	25	10.20	54.75	18.63
2	1999	36	11.30	54.70	20.65
3	2000	44	18.50	73.00	25.34
4	2001	38	24.10	91.25	26.41
5	2002	42	22.10	94.90	23.28
6	2003	52	30.10	109.50	27.48
7	2004	65	34.60	116.80	29.62
8	2005	78	32.10	131.40	24.42
9	2006	90	34.60	138.70	24.94
10	2007	110	38.10	156.95	24.27

Source: Primary Data.

In the mean while, there was a decline in 2001 which showed at 38 institutions a decline number in the over the immediate previous year. The company sold 10.20 lakh litres in 1998 to the institutions as compared to 38.10 lakh litres in 2007. There was a gradual growth in the institutional sales during the period expect in 2002 and 2005, which registered a decline over their immediate previous years. The proportion direct sales to institutions in the total sales are 18.63 per cent in 1998 which increased to 24.27 per cent in 2007 with relative ups and downs. It may be noted that in the institutions years the highest percentage of 29.62 per cent was registered in 2004. It can be concluded that there is an increase in the number of institutions and sale of milk directly during said period. Further to them their percentage share in the total sale of milk increased in the reference period with relative ups and downs. Joint reading of this table and preceding table reveals that the Nandi dairy is favourable placed in comparison to Vijaya dairy with regard to direct sales to institutions in absolute as well as relative terms.

Brand Names

Both dairies have high Grand equity for their products. Table-8.9 shows the Vijaya dairy markets milk and milk related products under single brand namely Vijaya. Similarly, Nandi dairy offers milk and milk related products under the brand name of Nandi. Each of these dairies has three product categories. The Vijaya dairy produces and sells twelve product variants whereas Nandi dairy thirteen. Details of product categories and product variants are always discussed in detail in the chapter five.

Trend of Sales through Different Channels of Distribution

Relative proportions of different channels in the total sales are discussed in this section. The details of methods of marketing channels of distribution used by Vijaya dairy and

Table 8.9. Product Line and Brand in Vijaya and Nandi Dairies

S. No.	Descriptions	Vijaya Dairy	Nandi Dairy
1	No. of. Product Variants	12	13
2	Product Categories	3	3
3	Brand	1	1

Source: Primary Data.

Nandi dairy during 1998-2007 are presented in Table 8.10. The Vijaya dairy distributes milk and by-products through direct and indirect methods. Sale through retail outlets and direct sale to institutions in bulk come under direct marketing. In the indirect marketing , dealer acts in between the dairy and consumers. The retail outlets are owned, managed, supervised and controlled by the dairy. The retail outlets sold milk and by-products worth Rs. 305.92 lakh in 1998 as against Rs. 817.61 lakh in 2007 which constitute 22.31 per cent and 20.21 per cent in the total sales respectively. During this period, there are fluctuations in the value of sales. For example, 2006 witnessed the highest sales of Rs. 978.13 lakh while the lowest sales of Rs. 280.10 lakh were recorded in 2000. The direct sale to institutions in large quantities increased from Rs. 355.83 lakh in 1998 to Rs. 915.11 lakh in 2007 in percentage terms, direct sale formed 25.95 per cent in 1998 as against 22.62 percent in 2007 with relative ups and downs. Thus, direct 48.28 per cent of sale for through direct marketing techniques while 42.84 per cent in 2007. the sale through dealers stood at Rs. 709.20 lakh in 1998 as compared to Rs. 3212.46 lakh in 2007 with relative fluctuations.

The Nandi dairy sold its products like Vijaya dairy, using direct and indirect sales methods. Of the total sales, retail outlets account for 5.57 per cent in 1998 as compared to 1037 per cent in 2007. The bulk sales to the institutions constituted 18.63 per cent in 1998 and moved upward with fluctuations and reached the highest percentage of. 65.34 per cent in 2007. In the recent four years there has a voluminous increase in direct sales to institutions which may be due to substantial

Table 8.10. Sales Through Retail Outlets, Dealers and Direct Selling in Vijaya and Nandi Dairies during 2007

(Rs. in lakhs)

Sl. No.	Particulars	Years									
		1998	1999	2000	2001	2002	2003	2004	2005	2006	2007
1	**Vijaya dairy** (a) Retail Outlet	305.92 (22.31)	310.52 (22,85)	280.10 (19.67)	291.00 (20.32)	517.21 (16.40)	334.37 (18.12)	334.37 (18.12)	334.37 (18.12)	978.13 (25.25)	817,61 (20.21)
	(b) Sales to Institutions	355.83 (25.95)	340.55 (25.06)	403.71 (28.35)	385.23 (26.97)	795.37 (25.22)	432.91 (23.46)	432.91 (23.46)	432.91 (123.46)	976.97 (25.22)	915.11 (22.62)
	(c) Dealer Sales	709.20 (51.72)	707.74 (52.08)	740.07 (51.77)	773.47 (54.01)	1840.51 (58.36)	1077.86 (58.41)	1077.86 (58.41)	1077.86 (58.41)	1907.07 (49.23)	2312.46 (57.16)
	Total	1371.23 (100.00)	1358.96 (100.00)	1424.04 (100.00)	1432.10 (100.00)	3153.73 (100.00)	1845.35 (100.00)	1845.35 (100.00)	1845.35 (100.00)	3873.81 (100.00)	4045.60 (100.00)
2	**Nandi dairy** (a) Retail Outlet Sales	49.84 (5.57)	54.79 (6.10)	57.13 (4.24)	128.59 (6.95)	53.25 (2.73)	106.36 (4.56)	89.47 (3.41)	295.65 (10.26)	306.49 (9.93)	376.49 (10.37)
	(b) Sales to Institutions	166.72 (18.63)	185.5 (20.65)	341.44 (25.34)	488.66 (26.41)	454.10 (23.28)	640.96 (27.48)	774.95 (66.94)	703.02 (65.62)	769.02 (65.09)	880.30 (65.34)
	(c) Dealer Sales	678.28 (75.79)	661.00 (73.58)	948.73 (70.41)	1232.86 (66.63)	1442.86 (73.97)	1584.67 (67.90)	1751.62 (29.61)	1889.40 (24.41)	2007.35 (24.93)	2370.33 (24.26)
	Total	894.95 (100.00)	898.35 (100.00)	1347.45 (100.00)	1850.32 (100.00)	1950.61 (100.00)	2332.46 (100.00)	2616.32 (100.00)	2878.87 (100.00)	3083.49 (100.00)	3627.14 (100.00)

Source: Primary Data.

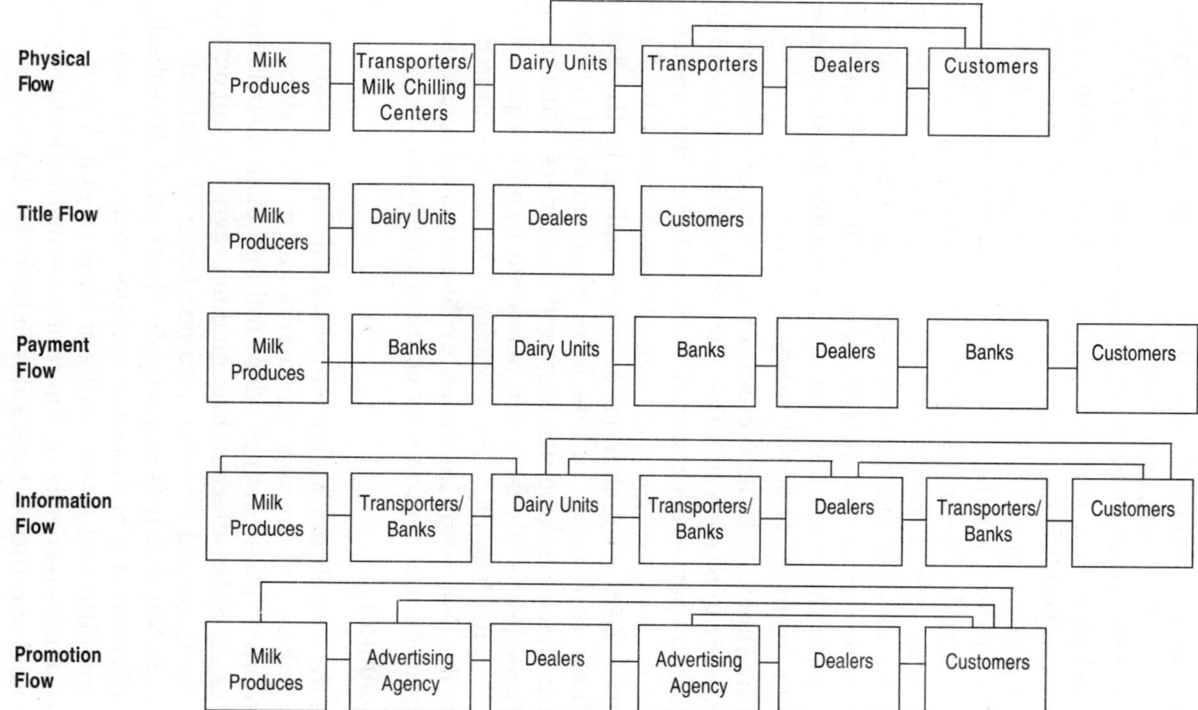

Fig. 8.1. Channel Functions and Flows for Dairy

increase in tie-ups. Three is a remarkable growth in the sale of milk by direct methods as against a substantial decline in the proportion of sales through dealers. This is based on the fact that the proportion of dealers sales through 75.79 per cent in 1998, it was whereas 24.26 per cent in 2007. It can be concluded that there is quantum important in sales through in both the dairies through direct marketing in the recent past as against distribution through indirect channel that is, dealers. Further, it can be observed that there is a substantial decline in the dealer sales in the Nandi dairy while a marginal increase in the Vijaya dairy. It may be further concluded that there is a substantial decrease in the share of dealer sales in Nandi dairy. A contrary situation is obtained in the Vijaya dairy but where is a marginal increase in the percentage of dealer sales.

Channel levels and channel players are shown in fig: 8.1. It can be seen that the diary and consumer and milk product are part every channel. The number of intermediary levels is the basis for the designating the length of the channel. Zero level channel or direct selling to the dairy products consumer. In one level channel stockists are intermediaries linking the dairy and consumers. Thus there are three levels of channel in operation in both Vijaya and Nandi units. The numbers of channel levels are limited in these dairies, resulting in no problems as to consumers' information and control over channel members.

Conclusion

Both the sample dairies have three channel levels 0-level, 1-level and 2-level through which milk and milk products are distributed. These defines adopt full and push distribution strategies. Sales affected through diary-owned retail outlets constitute around one-fifth and one-tenth of total sales in Vijay and Nandi dairies respectively. Major production of sales through come dealer-stockiest network come, which account for three-fifth and seven-tenths of total sales in Vijaya and Nandi dairies respectively. Share of sales institutional buyers is around one-fifth of total sales in both the dairies.

9 CHAPTER

SUMMARY OF FINDINGS, CONCLUSIONS AND SUGGESTIONS

This chapter is organized into four parts Part-A, Part-B, Part-C and Part-D. Part-A presents findings of the study, Part-B conclusions, Part-C suggestions and Part-D areas for further research.

PART-A: FINDINGS

Findings of the study are presented chapter-wise in this part.

CHAPTER-1

9.1 Share of livestock production in the country's GDP hovered round 6 per cent in late 1980s and mid 1990s. Which come to around 5 per cent in late 1990s and early 2000s. Thus, the share of this sector declined marginally.

9.2 Between 1951 and 2000 share of bovines in total livestock in India fell down from 68 per cent to 58 per cent, a decline of 10 percent points.

9.3 Annual growth rates of bovine population were negative in 1992-97 and 1997 to 2003 in contrasts positive rates registered in the earlier five year periods from 1951-56 to 1987-92.

9.4 India emerged as top milk producing country in the world in 2000 displaying America and its share in total world milk production was 18.54 per cent in 2004.

9.5 From 1951 to 2007 milk production in India registered continuous up-trend recording yearly positive percentage changes.

9.6 Per capita daily availability milk in India was 178 grams in 1992 which increased to 245 grams in 2007.

9.7 Emergence of Amul in Gujarat in 1946 to organize primary dairy cooperatives land mark event.

9.8 Yearly between 1990-91 and 2001-02 in were positive in the growth of societies ranging from the lowest of 0.67 per cent to the highest of 15.56 per cent and the growth of membership ranging from the lowest of 1.77 per cent to the highest of 33.06 per cent.

9.9 OF programme assisted by WFP and implemented in three phases between 1970 and 2002, had creditable achievements in milk procurement, processing capacity created milk marketing and provision of technical inputs?

9.10 Share of Animal Husbandry and dairying in total plan outlay was very low at 0.11 per cent in first five year plan which reached maximum of 1.47 per cent in fourth five-year plan.

9.11 Share of primary sector in AP state GDP and employment was 52 per cent and 71 per cent respectively in 1950-51, which came down to 32 per cent to 70 per cent in 1999-2000. Through this sector witnessed a declining share in GDP, it had dominant share in total employment 9.12 per cent in A P witnessed significant growth in milk production from 1.68 million tonnes in 1991 to 3.85 million tonnes in 2001 and in per capita daily milk availability from 108 grams in 1971 to 147 grams in 2001.

9.12 AP witnessed significant growth in milk production from 1.68 million tonnes in 1991 to 3.85 million tonnes in 2001 and in per capita daily milk availability from 108 grams in 1971 to 147 grams in 2001.

9.13 Share of female cattle over 3 years in total cattle was 35.79 per cent in 1985 and 35.30 per cent in 2002, indicating no change. Share of female buffaloes over three years in total buffaloes was 49.69 per cent in 1985 and 50.04 per cent in 2002, revealing no change.

9.14 Milk production in AP was 4.47 million tonnes in 1997-98 which increased to 7.94 million tonnes in 2006-07, registering LGRs of 4.33 per cent.

9.15 Per capita availability of milk was 167 of grams per day in 1997-98 which increased to 269 grams per day in 2006-07, registering CGR of 6.21 per cent.

9.16 Milk procurement in A P was 3112.42 lakh liters in 1997-98 which increased to 5445.45 lakh liters recording in 2007.

9.17 Milk sales in A P were 2273.00 lakh liters in 1997-98 which increased to 4568.54 lakh liters registered in 2007, CGR of 6.69 per cent.

CHAPTER-3

9.18 Vijaya dairy cooperative established in Kurnool district by APPDDC in 1974 collects milk from 36,000 milk producers in 566 villages through 256 MPCs and 463 MPACs Nandi dairy a private sector dairy functioning in Kurnool district was established in 1997 by Nandi group of industries. It covers 16,845 milk producers in 324 villages through 324 MPCs/ MPACs.

9.19 Annual procurement of milk by Vijaya dairy was 92 lakh litres in 1998 which increased to 322.18 lakh liters in 2007, registering more than three-fold increase. In contrast Nandi dairy 14 lakh liters in 1998 which increased to 153.66 lakh liters in 2007, 11-time increase.

9.20 Price per litre offered in both the dairies is based on fat percentage in milk, higher the price. Litre price is increasing is related to SNF content in the milk and average milk procurement price per litre are 8 per cent and Rs. 15.60 respectively in both dairies.

9.21 Gold milk of Vijaya dairy is sold at premium price of Rs. 20 per litre whereas its skimmed milk is priced low at Rs. 12 per litre.

9.22 All milk products uniformly have one day shelf life. Skimmed milk powder of Vijaya dairy *ghee* of both the dairies, butter Vijaya dairy has shelf lives of 1 year, 6 months and 90 days respectively. Sweet products of Nandi dairy have shelf life of 5 days.

9.23 Annual milk sales of Vijaya dairy more 77. 62 lakh liters in 1998 and 1995 lakh liters in 2007, registering more than two-fold increase. In the corresponding years, Nandi dairy sold 54.75 lakh liters and 156.95 lakh liters, recording abut threefold increase.

9.24 Between 1998 and 2007, sale of panneer, butter milk, *doodhpeda* skimmed milk powder, sterilized flavoured milk, butter and *ghee* of Vijaya dairy registered 1391.33, 299.52, 106.25, 82.35, 24.56, 4.51 and -40.78 percent Increased respectively.

9.25 All sweet products of Nandi dairy are of equal importance in terms of quantities sold. Their respective percentage shares in total quantity sold in 2007 varied narrowly from 18.13 per cent share of *kalakhan* to 15.29 per cent share of *khoya*.

9.26 Between 1998 and 2008, quantified sold with respect to ghee and butter milk registered more than two-fold and about four-fold increases respectively in Vijaya dairy.

CHAPTER-4

9.27 Daily average milk procurement per society in Vijaya dairy was 52.55 litres in 1998 and 122.76 liters in 2007, recording more than two-fold increase in contrast in three corresponding years; Nandi's average was 29 liters and 104 liters, registering three-fold increase.

9.28 Annual milk production was 92.08 lack liters in 1998 and 322.18 lakh litres in 2007 in Vijaya dairy, registering more than three-fold increase. In comparison, annual production was 14 liters and 153.66 lack litres in corresponding years in Nandi dairy, registering more than ten-fold increase.

9.29 Daily milk procurement in lean season expressed as percentage of daily procurement in flush season was lowest at 61 per cent in 2005 and 2007 and is highest at 82 per cent in 2001 in Vijaya dairy, as compared to the lowest of 61 per cent in 2001 and highest of 72 per cent in 2004 in Nandi dairy. C.V. vales of average dairy procurement in flush and lean seasons were 41.1 per cent and 35.51 per cent respectively in Vijaya dairy, and 71.68 per cent and 70.38 percent in Nandi dairy.

9.30 During 2007, Vijaya dairy achieved chilling capacity utilization of 40.72 per cent as compared 71.31 per cent capacity utilization archived by Nandi dairy

CHAPTER-5

9.31 Both dairies have three product lines each. In terms of number of products and by product lines are major product lines in Vijaya dairy. In comparison in Nandi dairy milk product and sweet product lines are major once.

9.32 In terms of sales the share of milk product and by products was 95.39 and 4.36 per cent of total sales respectively in 1997 in Vijaya dairy. In comparison milk products and sweet products accounted for 86.87 and 11.83 per cent of total sales respectively in Nandi dairy in 2007.

9.33 Product depth in Vijaya dairy increased from two products in 1977 to 12 products in 2007. In comparative analysis product depth of Nandi dairy had three products in 1997 and 13 products in 2000.

9.34 Products of both dairies are sold either in tetra packs or cardboard packs. Products are sold under 'Vijaya' and 'Nandi' brand names.

9.35 Both dairies adopt ingredient branding practices indicating fat and SNF contents in respective products sold by them.

9.36 Up-market stretch was adopted by both dairies with regard to only milk products.

9.37 Products classified alternately core and staple products accounted for 90.40 per cent and 9.60 per cent in total sales respectively in Vijaya dairy and 83.4 per cent and 16.57 percent in total sales respectively in Nandi dairy.

9.38 LGRs of all milk products basic product and 2 out of 6 by-products are significant in Vijaya dairy in the period from 1998 to 2007 dairies. In comparison LGRs of all products of Nandi dairy are statistically significant in the period from 1998 to 2007.

CHAPTER-6

9.39 Maximum profit was the pricing objective pursued in both the dairy in respect of majority of products.

9.40 'Statistically techniques' were used for demand estimation of most of the products in both the dairies, followed by price experiments and surveys.

9.41 Market-determined pricing 'value pricing' and 'going rate pricing' are the pricing methods adopted for large number of products both dairies.

9.42 Both dairies offer discounts of various kinds to dealers ranging from 8 to 10 per cent, to institutional buyers ranging from 6 to 8 per cent and individual consumers ranging from 5 to 7 per cent to promote sales and induce prompt payments.

CHAPTER-7

9.43 Taking actual number of advertisement programes as against the ideal number of programmes the per cent score checklist was maximum 100 per cent in Vijaya dairy as against 92.3 per cent in Nandi dairy.

9.44 Checklist score of sales promotion programmes was highest in Nandi dairy at 66 per cent as against 30.76 per cent in Vijaya dairy.

9.45 Checklist score of events/experiences was least in Vijaya dairy at 0 per cent as against 37.50 per cent in Nandi dairy.

9.46 Check list score of public relations was 60 per cent each in both dairies.

9.47 With regard to direct marketing practices, the check list score low at 12.50 for both the dairies.

9.48 In milk order of six communication mix elements, advertising and personal selling are prominent in sample dairies.

9.49 Checklist score of methods used for determining total communication budget was 75 per cent in Vijaya dairy and 50 per cent in Nandi dairy.

9.50 Average total communication budgets for the period from 1998 to 2007 was Rs. 2.18 lakh in Vijaya dairy and Rs. 2.39 lakh in Nandi dairy. LGRs of communication budgets were 13.54 per cent 1.78 per cent in Vijaya dairy and Nandi dairy respectively. Which were statistically significant?

9.51 LGRs were of higher magnitude for both sales and communication budgets in Nandi dairy than in Vijaya dairy.

9.52 LGRs of advertising budget were of higher magnitude in Vijaya dairy (12.66 %) than in Nandi dairy (11.61 %).

9.53 In sales promotion tools adopted, point of purchase displaces and demonstrations were used for majority of products in both Vijaya dairy and Nandi dairy.

9.54 Advertising budget utilization as percentage of budget allocation ranges from the lowest 92.28 per cent in 2003 and the highest of 116 per cent in 1999 in Vijaya dairy, whereas in Nandi dairy the lowest utilization was 92.50 per cent in 2000 and the highest 106.66 per cent in 1999.

9.55 Between 1998 and 2007 the average spending on advertisement was highest at Rs. 1.17 lakh in news papers, and lowest at Rs. o.28 lakh in magazines in Vijaya dairy. In Nandi dairy the average highest spending was Rs 1.21 lakh in newspapers and lowest being 0.54 lakh on television. LGRs of spending in advertisement in different media are significant except for broachers/pamphlets. In LGRs of spending on advertisement in Nandi dairy in different media are significant except for magazines.

CHAPTER-8

9.56 For distribution of their products, both the dairies have 3 channels in 0-level channel they sell directly to consumers, in 1-level channel dealers are involved in between dairies and consumers and in 2-level channel dealers and stockiest are intermediaries between dairies and consumers.

9.57 Sales through dairy's sales outlets as percentage of total sales ranged from the lowest of 15.67 per cent in 2004 to 25.88 per cent 2005 in Vijaya dairy and the minimum being 2.73 per cent in 2002 and maximum being 10.38 per cent in 2007 Nandi dairy.

9.58 Dealers-effected sales as percentage of total sales was lowest at 49.23 per cent in 2006 and highest at 59.05 per cent in 2005 in Vijaya dairy, as against the lowest of 65.10 per cent in 2006 and highest of 75.79 per cent in 1998 in Nandi dairy.

9.59 Sales to institutional buyers constitute significant proportion of total sales in both the dairies between 1998 and 2007 it proportion was lowest at 14.60 per cent in 2005 and highest at 28.35 per cent in 2000 in Nandi dairy, as against the lowest at 18.03 per cent in 1998 and the lowest at 29.62 per cent in 2003 in Nandi dairy.

9.60 Sales to institutional buyers were by far the largest proportion of total sales in Nandi dairy with per cent share of 65.34 per cent in 2007, as against 57.16 per cent share dealer affected sales in 2007 in Vijaya dairy.

PART-B : CONCLUSIONS

9.61 Livestock was marginal sector in its contribution to GDP, but a major sector in its contribution to employment.

9.62 India is the top milk producing country of the world closely followed by U.S.A.

9.63 Since 1951 till date India achieved remarkable growth in milk production and per capita daily availability of milk.

9.64 Reflection of Anand pattern of organization of dairy cooperatives and OF programmes I, II and III brought white revolution in India.

9.65 Plan allocations to Animal Husbandry and dairying were woefully low, not reflecting its importance Indian economy in general and its rural economy in particular.

9.66 Proportion of female buffaloes over three years of age was more as compared to female cattle in A P.

9.67 Both in milk procurement and sales A P has enable record with LGRs of 4.22 and 3.79 per cent respectively.

9.68 Increase in milk procurement was more in Nandi dairy than in Vijaya dairy.

9.69 All milk products have shelf life of one day as compared to 5-day shelf life of sweet products, 6 months shelf life of ghee and 1-year shelf life of skimmed milk powder.

9.70 Sales of milk recorded two-fold increase in Vijaya dairy and three-fold increase in Nandi dairy, later having a better milk sales record.

9.71 Daily average milk procurement per society was more in Vijaya dairy than in Nandi dairy.

9.72 Wide variations were profound in milk procurement in flush and lean seasons.

9.73 Milk product line is the dominant product line in total sales value in both the dairies.

9.74 Up-market stretch was adopted by both dairies with record to milk sales only.

9.75 Both dairies pursue maximum profit pricing objective, use 'statistical techniques' for demand estimation majority of products, adopt 'market-determined' 'value pricing' and 'going rate pricing' methods and offer quantity and cash discounts of higher magnitudes to dealers and institutional buyers to individual consumers.

9.76 Going by the checklist percentage scores Vijaya dairy had better promotional practices in advertisement programmes, whereas Nandi dairy had better practices in sales promotions programmes, events and experiences.

9.77 Communication budgets recorded impressive growth in both the dairies.

9.78 Major media of advertisement was news papers in both the dairies.

9.79 Dealer-effected sales were the largest component of total sales followed by sales to institutional buyers in both the dairies.

PART C : SUGGESTIONS

9.80 There is a need for increasing public spending in dairy through increased budget outlays buy central and state governments.

9.81 Cooperative dairying needs to be further strengthened to provide augmented in put supplies to improve the milk yields.

9.82 Per capita availability of milk per day needs to be augmented through better procurement, process and marketing practices, strengthen cooperative dairies and initiating FDI in private sector dairy industry.

9.83 Value added products should be introduced having larger shelf lives.

9.84 Tailor-made incentive structure should be evolved by the dairies to get increased supply of milk from producers which are otherwise routed to through players in unorganized markets.

9.85 To achieve higher levels of capacity utilization in chilling centers, dairies have to work out both in-found and out-found logistics.

9.86 Down-market stretch practices should be adopted by the dairies, offering products in small packages in mass markets.

9.87 Sales promotions efforts which are nominal at present should be strengthened.

9.88 There should be proper evaluation of communication mix so as to assess its impact on sales.

9.89 To keep tuned with revolutionary changes in retailing, initiatives should be taken by the dairies to sell milk through cans through super markets.

9.90 Dealers network should be strengthened, offering them proper discount and other incentives.

PART-D : AREAS FOR FURTHER RESEARCH

Further research in marketing area may be under taken with focus on the following :

9.91 Dealer motivation, functions and sales performance is one such potential area.

9.92 Comparative study may be taken to study consumer attitudes towards various branded dairy products.

9.93 'Brand equity' of products of dairy units operating in private and cooperative sectors may be another potential area of research.

BIBLIOGRAPHY

BOOKS

Acharya, S.S.and Yadav, R.K. *Production and Marketing of Milk and Milk Products in India* New Delhi, Mittal Publications, 1992.

Allen, Agnes *The Co-operative Story, Cooperative Union Limited,* Holyoake, Manchester, 1963.

Avery, Devid V. *Market Structure, Conduct and Performance at the Midwest Dairy Industries*, un-published Ph.D. thesis, Madison University at Wisconsin, 1966.

Bagavathi and R.S.N. Pillai, *Modern Marketing Principles and Practices,* Chand & Company Limited, New Delhi, 1987.

Bedi, R.D. 1958. *Theory, History and Practice of Co-operation,* Meerut Loyal Department.

Bhanja, S.K. and Venkatadri, S. *Milch Cattle in IRDP*, Hyderabad, National Institute of Rural Development, 1988.

Cole, H.H. and Manager Rouning. 1980. *Animal Agriculture*, W.H. Freeman and Company, Sanfarancisco.

Goel, B.B. *Cooperative Management and Administration*, New Delhi, Deep and Deep Publications, 1984.

Gupta. P.R. *Dairy India,* New Delhi, Rekha Printers., 1997.

Hajela, 1987. *Principles, Problems and Practice of Co-operation,* Shivalal Agarwala and Company, Agra.

Jain, M.M. *Growth Pattern of Dairy Sub-Sector in Rajasthan: A Study of Milk Producers Cooperatives,* Bombay, Himalaya Publishing House, 1986.

Jawana, Ram *Management of Dairy Enterprises,* Jaipur, Kuber Associates And Publishers, 1987.

Judkins, Henry H. *et. al. Milk Production and Processing,* Wiley Eastern India Limited, Newyork, 1960.

Kamath, M.V., 1989. *Management Kurien Style – The Story of the White Revolution* Konark Publication Pvt. Ltd, New Delhi.

Karthikeyan, S. *A Study of Production and Marketing of Milk in Mayuram Area*, Annamalai University Press, Annamalai Nagar, 1980.

Khan, M.Y. and Jain, P.K. 2001, *Financial Management Text and Problems,* Tata McGraw Hill Publishing Company Limited, New Delhi.

Khurody D.N. 1974. *Dairying in India – A Review,* Asia Publishing House, Delhi.

Kotler, Philip and Kevin Lane Keller. *Marketing Management*, Pearson and Prentile Hill, Delhi, 2006. Agarwal, V.K. Marketing of Dairy Products in Western Uttar Pradesh, Himalaya Publishing House, Bombay,1976.

Lakshman, T.K. and Narayan, B.K. *Rural Development in India: A Multi Dimensional Analysis*, Bombay, Himalaya Publishing House, 1984.

Latif, T.A.A. *Marketing Management*, New Delhi, Deep and Deep Publications, 1983.

Madan Mohan, C. *Dairy Management in India*, New Delhi, Mittal Publications, 1989.

Mamoria C.B. and R.D. Saksena, *Co-operation in Foreign Lands*, Kitab Mahal Private Limited, Delhi, 1963.

Mamoria, C.B. and Tripathi, B.B. *Agricultural Problems of India*, New Delhi, Kitab Mahal Publications. 1985.

Mascarenhas, *A Strategy for Rural Development and Dairy Co-operatives in India,* Sage Publications, Bombay, 1988.

Mirchandan, *Animal Husbandry*, Allied Publishers, Bombay, 1979.

Misra, S.N. *Livestock Planning in India*, New Delhi, Vikas Publications, 1978.

Mitra M., *Women's work: Gains Analysis of Women's Labour in Dairy Production*, Sage Publication, New Delhi, 1987.

Muniraj, *Farm Finance for Development,* Oxford and IBH Publishing Company Private Ltd., New Delhi, 1987.

Patel A.S., *Cooperative Dairying and Rural Development: A Case Study of Amul,* Oxford University Press, New Delhi, 1988.

Pillai, Manikya Vasagam N. *A Study on Resource use Efficiency in Milk Production in Parimbikulam Aliyar Project Region, Tamilnadu*, un-published Dissertation, Coimbatore, 1976.

Prasad, Durga P. *Cooperative Dairying and Rural Development: A Case Study of Villages in Guntur District of Andhra Pradesh*,

Un-published Ph.D. thesis. Sardar Patel University, Vallabha Vidyanagar, Gujarat, 1986.

Regeena, S. "Small Holder Livestock Production in Home Gardens of South Kerala, Livestock in Different Farming Systems in India, 2002, pp. 120-131. Advance Publishing Concept, New Delhi.

Shah, Dilip 1986. *An Economic Analysis of the Co-operative Dairy Industry,* Somaya Publishers, Bombay.

Shanthi, George 1985. *Operation Flood – An Appraisal of Current Indian Dairy Policy,* Oxford University Press, New Delhi.

Subbulekshmi, B., *Profitability of Dairy Farming: A Case Study*, Department of Economics, Arul Anandar College, Madurai, 1998.

Sukhatme, P.V., *Feeding India's Growing Millions*, Asia Publishing House. Bombay,1978.

Sukumar, D.E., *Outlines of Dairy Technology* Oxford University Press, Bombay 1980.

Warner, *Dairying in India,* Macmillan & co., New York, 1951.

Warner, James N. *Principles of Dairy Processing,* Wiley Eastern India Limited,New Delhi, 1994.

JOURNALS

Datta, T.N. and Murlidhar, J.B. "Resource Endowment and Milk Production Economy: Sine Reflection on West Rajasthan District," *Indian Journal of Agricultural Economics*, Vol. 50, No. 3, 1995, pp. 326-328.

Gaddi, G.M. and kunnal, L.B. "Sources of Output Growth in New Milk Production Technology and Some Implications to Returns on Research Investment", *Indian Journal of Agricultural Economics*, Vol. 51, No. 3,1996, pp.389-395.

Gandhi, R.S. and Singh, A. "Crossbreeding of Cattle for Enhancing Milk Production under Sustainable Livestock System in India," *Indian Dairyman*, Vol. 56, No. 7, 2004, pp.51-57.

Gaurasha, A.K. "Comparative Economics of Milk Production in Urban and Rural Areas of Madhya Pradesh", *Indian Journal of Agricultural Economics*, Vol.50, No.3, 1995, pp. 365-366.

Goplakurup, P.T. "Clean Milk Production and Marketing of Milk from the Farmers' Perspective", *Indian Dairyman*, Vol.54, No. 2, 2002, pp.92-95.

Khanna, R.S. Sustainable Development of the Dairy Industry in

Asia Challenges and Opportunities, *Indian Dairyman*, Vol. 57, No. 2, 2005, pp.21-29.

Khatkar, R.K. et.al. Livestock as a Tool of Diversification and Risk Management, *Agricultural Research Review*, Vol. 13, No. 2, 2000, p.201.

Khodaskar, R.D. "Economics of Dairy Enterprise with Crossbred Cows: A Case Study of Marginal Farmers of Ranjani Village (Pune)," Indian journal of Agricultural Economics, Vol.50,No.3,1995,p.359.

Kumar, Anjani and Gupta, J.N. "Impact of Crossbreeding Technology of Cattle in Bihar: A Micro Perspective," *Agricultural Economics Research Review*, Vol. 13, No. 2,2000, pp.169-177.

Lalwani, Mahesh and Mahesh "Technology Change in Indian Dairy Farming Sector: Distribution and Decomposition of Output Gains", *Indian Journal of Agricultural Economics*, Vol. 44, No. 1, 1989, pp. 55-66.

Manbhekar, M.V. et. al. "Relative Economics of Milk Production for Local vis-à-vis Crossbred Cow: A Study in the Vicinity of Akola City (Maharastra), *Indian Journal of Agricultural Economics*, Vol. 50, No.3, 1995, p.364.

Mehta, Pradeep. S. "Indian Dairy Sector: the Challenges Ahead", *Indian Dairyman*, Vol. 56, No.10, 2004, pp.152-158.

Naik, Dibakar and Mohanty, Binod Ch. "Economics of Milk Production with special reference to Resource use in the Existing Market Environment of Orissa", *Indian Journal of Agricultural Economics*, Vol. 50, No.3, 1995, P. 336.

Natchinmuthu, K. and Kumar, Ram "Potentiality of Increasing Farm Income and its Complimentarily with Dairy Enterprise in Cotton Zone of Surat District in Gujarat State", *Agricultural Situation in India,* Vol. LIII, No. 11, 1997, pp. 751-757.

Radha, Y. *et. al.* "Study on Income and Employment Generation in Agriculture Based Livestock Farming System, *Agricultural Economics Research Review*, Vol. 13, No. 2, 2000, p. 211.

Ramesh Babu, P. "Trends Patterns and Effects of Diffusion and Adoption of Crossbreeding Technology: An Assessment in the Context of Kerala", *Indian Journal of Agricultural Economics*, Vol. 50, No.3, 1995, pp. 294-298.

Rathi, Deepa K. *et. al.* "Sustainable Employment Generation through

Livestock Enterprise: A Case Study of Bhind District"., *Agricultural Economics Research Review*, Vol. 13, No. 2, 2000, pp. 211-212.

Reddy, K.P. "Initiatives for Achieving Excellence: Indian Dairy Industry Scenario," *Indian Dairyman*, Vol. 56, No. 10, 2004, PP. 44-49.

Rout, Niranjan and Tripathy, "Economics of Milk Marketing in Khurda District of Orissa," *Indian Journal of Agricultural Economics,* Vol.50, No.3, 1995, p.336.

Sankar and Ravi. Policy and Regulations: Influencing Milk Production," *Indian Dairyman*, Vol. 55, No.3, 2003. PP. 96-100.

Shah and Deepak. "Making Milk Production Viable to Farmers-An Analysis of Co-operative Approach," *Agricultural Situation in Indian*, Vol. LIV, No.6, 1997, pp.375-380.

Sharma, Vijay Paul and Sharma, Pritee. "Implications of Agricultural Trade Liberalization for Indian Dairy Industry," *Agricultural Economics Research Review*, Vol. 13, No.2, 2000, pp.183-198.

Shukla, D.S. *et. al.,* "Impact of Operation Flood Programme on the Economy of Rural Milk Producers in Districts of Kanpur-Dehat," Indian Journal of Agricultural Economics, Vol. 50, No.3, 1995, PP.3710372.

Singh, Rajvir and Sainiamrik, S. "Integration of Improved Technology of Crop and Milk Production for Increasing Income and Employment," *Agricultural Situation in Indian*, Vol. XLIII, No.9, 1998, PP.751-755.

Singh, R.P. "Contribution of Livestock and Crop Enterprises to the Economy of Tribal Farmers", *Indian Dairyman*, Vol. 50, No.3, 1995, pp. 363-364.

Soundarapandian, M."Impact of Milch Animals Scheme of IRDP on Agricultural laborers: A Case Study of Kamarajar Distirct in Tamilnadu," *Indian Journal of Agricultural Economics*, Vol. 50, No.3, 1995, p. 324.

Thamas, George and Mani, K.P. "Import Service Management of Dairy Co-operatives: A Farmer Oriented Evaluation," *Indian Journal of Agricultural Economics*, Vol. 50, No.3, 1995, p. 377.

Vashist, G.D. and Katha, Pradeep. "A Comparative Economic Analysis of Milk Production for Different Milch Animals in

Himachal Pradesh", *Agricultural Situation in India*, Vol. XLIII, No.2, 1988, pp. 133-138.

REPORTS AND DISSERTATIONS

George Mergos and Roger slade, "Dairy Development and Milk Cooperatives: The Effects of Dairy Project in India", The Work Bank, Washington, 1987.

Government of India, "Basic Animal Husbandry Statistics", Department of Animal Husbandry and Dairy, Ministry of Agriculture, 2002.

Government of India, "Report of the National Commission on Agriculture", Vol.7(1) Ministry of Agriculture and Irrigation, New Delhi, 1976.

Indian Dairy Association. "Dairy Industry Conference", XVII, Special Number: Dairying in India, New Delhi, 1988.

Ranganathan K and Sinha M.N. "A study of Differential Participation of Rural Women of Southern States in Dairy Farm Sector", Annual Report PP.94-95, Southern Regional Station, Natinal Dairy Research Institute, Bangalore, 1995.

INDEX

A

Aarey Milk Colony, 11

All India Rural credit Review Committee (AIRCRC), 34

AMUL, 11

Anand pattern, 12-13

Andhra Pradesh Dairy Development Corporation (APDDC), 1

Annual plans (1990-92), 45

Artificial Insemination (AI), 13

Artificial Insemination Centres (AICs), 89

B

Balaji, 126

Branded dairy products, 216

Bureau of Indian Standards (BIS), 96

C

Co-operative Societies Act, 1912, 11

D

Dairy Cooperative Societies (DCS), 71

Dairy development in Andhra Pradesh, 36-56

- A.P. State domestic product and its sectoral shares, 37-38
- Annual Plans (1990-92), 45
- bovine population, 50-53
- dairy development during plan periods, 43
- dairy development under OF, 46-49
- development of dairy infrastructure, 49-50
- formation of dairy development co-operative federation, 40
- formation of the APDDC, 39-40
- growth and structure of dairying, 38-39
- membership of federation, 41-43
- milk production and per capita availability of milk, 53-56
- procurement and sale of milk, 56-58
- significance of dairying in Andhra Pradesh, 36-37

Dairy development in India

- Anand pattern, 12-13
- challenges ahead in dairy sector in the country, 30-31
- co-operative dairying, 10
- dairy development during
  - first two decades of planning 1951-71, 22-23
  - plan periods, 21-22
- dairying in post-reform period, 27

development, 11-12
establishment of cattle colonies and milk schemes, 26-27
fodder development, 22
global diary scenario, 7-9
growth of crossbreed cows vis-à-vis indigenous cows, 3-6
information and development research, 30
intensive cattle development project, 26
inter-State variations in milk production and yield, 6-7
introduction of milk and milk products order, 27-28
key village scheme, 25-26
milk consumption pattern in
India, 34-35
other countries, 34-35
operation flood programmes, 14-19
operation flood-I, 15-17
operation flood-II, 17-18
operation flood-III, 18-19
origin, 11
perspective 2010, 29
present status of operation flood programme, 19-20
production enhancement, 29
prospects of dairy industry in India, 35-36
quality assurance programmes, 30
replication of Anand pattern, 13-14
responses to meet the challenges, 31-34
significance of dairy farming in India, 1-3
steady growth of diary sector, 9-10
strengthening co-operative business, 29
technology mission on dairy development, 23-25
Uruguay round agreement on agriculture, 28
Dairy India Year Book, 80
Dealer, motivation, functions and sales performance, 216
Delhi Milk Scheme (DMS), 21
Department of Animal Husbandry and Dairying (DAHD), 34
Doodhpeda, 96, 99, 111

E

Eighth five-year plan (1992-97), 45
European Economic Community (EEC), 18

F

Fifth Five-year Plan (1974-78), 44
Findings, 205
Five-year Plans, 1
Fourth five-year plan (1969-74), 44

G

Government of India, 17

I

Indian Council of Market and Research (ICMR), 9, 35
Indian Dairy Association, 17
Indian Dairy Corporation (IDC), 40
Indian Immunological Technology Mission (IITM), 20

Integrated Rural Development Programme (IRDP), 2, 24
Intensive Cattle Development Programme (ICDP), 22
Intensive Cattle Development Scheme, 75
Introduction, 1-62

K

Kalaghan, 111
Kamadhenu, 126
Katara Co-operative dairy Limited at Allahabad, 11
Khoya, 111, 208

M

Madhya Pradesh, 17
Madras Milk Supply Union, 11
Mandi Milk Supply Scheme of Himachal Pradesh, 67
Milk and Milk Products Order (MMPO) in 1992, 27
Milk Cooperative Societies/Milk Collection Agents (MCS/ MCAs), 116
Ministry of Agricultures in 1991, 34
Multi-National Companies (MNCs), 2

N

Nandi dairy, 81, 100-115, 209
Nandi Milk Products Pvt. Ltd., 102
National Animal Health and Production Information System, 34
National Commission on agriculture, 22
National Council of Applied Economic Research (NCAER), 71
National Dairy Development Board (NDDB), 13, 24
National Dairy Research Institute, 75
National Milk Grid (NMG), 19
NGOs, 29

O

Objectives and research methodology of the study, 77
  data sources, 80
  need for and significance of the study, 78-79
  objectives of the study, 79
  organisation of the book, 81
  sample design, 80
  scope and limitations, 80-81
  statement of the problem, 77-78
  tools of analysis, 80
Operation Flood (OF), 1, 70
Operation Flood Programme, 9, 14-19

P

Patel, Sardar, 12
Phule, Mahatma, 75
Plakova, 111
Pricing objectives and methods of dairies, 155-165
  demands estimations techniques, 157-160
  introduction, 155
  price discount and allowances, 164-165
  pricing methods, 160-163
  pricing objectives, 155-157
Product related marketing practices, 128-154

distribution of sales by-product lines, 147-154

ingredient branding, 139-141

introduction, 128

product depths in samples dairies, 134-136

product hierarchy, 129-131

product-line sales in sample dairy, 131-134

product of sample dairy units, 128-129

stretching of packaging and labelling in sample dairies, 136-138

trend of sales of core and staple dairy products, 143-147

up and down stretching of product lines, 141-143

Profile of Nandi dairy, 100-115

beginning of Nandi group of industries, 100-102

channels of distribution of Nandi dairy, 109

milk procurement by Nandi dairy, 103-104

organization chart of Nandi dairy, 111-115

origin and growth of the Nandi dairy, 102-103

procurement prices, processing and sales in Nandi dairy, 104-109

sales of milk products, by-products and products prices of Nandi dairy, 109-111

Profile of sample dairies, 87-115

climate conditions and revenue administration set-up, 87-88

geographical location of Kurnool district, 87

infrastructure for dairy development, 88

introduction, 87

Profile of Vijaya dairy, 88-93

contents and shelf life to milk and milk products, 93-95

feed, 90

fertility campaigns, 90

insurance, 90

milk procurement and prices are discussed in this section, 91-93

origin of KDMPMACU, 88-89

others services, 91

price fat and SNF contents and shelf life to milk products and milk by products of Vijaya dairy, 95-97

provision of de-worming drugs, 90

sales of milk, milk products and by-products in Vijaya dairy, 97-100

services of KDMPMACU Ltd., 89

urea treatment, 91

vaccination, 90

Promotion practices in sample dairy units, 166-190

advertising budgets and sales, 179-183

advertising programmes in sample dairy units, 166-167

communication budgets and sales, 173-178

direct marketing, 172

events/experiences programs, 169-170
introduction, 166
media-wise advertising budgets, 184-190
personal selling, 171-172
public relations, 170-171
rank order of communication mix, 173
sales promotion, 183-184
sales promotion programmes, 167-169

R

Rayalaseema, 37
Review of literature and research design, 63-86
Acharya, 69
Bhattacharya, 76
Grill and Grill, 69
introduction, 63
Kaur, Mahindra, 64
Misra, 68
Mohan, Madan, 76
Patel, 64
Prabaharan, 65
Ramachandran, 68
Ravikiran, 68
studies on
bovine population and milk yields, 73
capacity utilization and production, 74
cost structure and returns in dairying, 74-77
economics of dairying, 68-71
marketing of dairy products, 63-68
the consumption and availability of milk, 73-74
women participation in dairying, 71-72
Subramanyam, 72
River Krishna, 87
Royal Commission on Agriculture, 3

S

Sales and distribution practices of dairies, 191-204
brand names, 200
distribution through dealer networks, 195-197
instructional vis-à-vis total sales, 198-200
introduction, 191
marketing channel systems and strategies, 191-193
sales through dairy owned retail outlets, 193-195
trends of sales through different channels of distribution, 200-204
Sastri, Lal Bahadur, 13
Second Five-year Plan (1956-61), 43
Sectoral shares, 37-38
Seventh Five-year Plan (1985-90), 45
Sixth Five-year Plan (1980-85), 44-45
Skim Milk Powder (SMP), 15
Small Farmers Development agency (SFDA), 2
State Milk Marketing Federations (SMMF), 30

Suggestions, 215

T

Tamil Nadu Dairy Development Corporation, 67

Technology Mission on Dairy Development (TMDD), 24

Telangana, 37

Tenth Five-year Plan, 34

Third Five-year Plan, 43-44

Trends in milk procurement in sample dairies, 116-127

- capacity utilization in milk chilling centers of vijaya and Nandi dairies, 124-127
- introduction, 116
- lean and peak seasonal variations in milk procurement in
  - Vijaya dairy, 119-122
  - Nandi dairy, 122-124
- milk procurement and production in Vijaya dairy, 116-117
- trends in milk procurement and production in Nandi dairy, 118-119

U

UNICEF, 40

United Kingdom, 7

United States, 35

Uruguay round agreement on agriculture, 28

U.S.A., 213

V

Vijaya Co-operative Milk Producers Union Limited, 81, 88-93, 116-117, 130, 210

W

White Revolution, 9

World Bank, 17

World Food Programme (WFP), 14

World Trade Organization (WTO), 1, 2, 28